APPLIED CHEMISTRY

APPLIED CHEMISTRY

By

G. Singh

Reader

Dept. of Chemistry

C.C.S. University

Meerut (U.P.)

(India)

D P H

DISCOVERY PUBLISHING HOUSE PVT. LTD.

NEW DELHI-110 002

First Published – 2009

Reprinted – 2026

ISBN: 978-81-8356-439-7

Applied Chemistry

Published by:

DISCOVERY PUBLISHING HOUSE
4383/4B, Ansari Road, Darya Ganj
New Delhi-110 002 (India)
Phone: +91-11-23279245; 23253475; 43596065
Mobile: +91 9811179893 / +91 9871656464
E-mail: discoverybooksindia@gmail.com
orderdphbooks@gmail.com
namitwasan9@gmail.com
web: www.discoverypublishinggroup.com

Printed at:
Infinity Imaging Systems
Delhi

Contents

Preface

Applied Chemistry is particularly in vogue, in areas of Chemical Technology, Engineering and Pharmacy. Other areas for study include Organic Synthesis, Catalysts, Surface Chemistry, Surfactants, Pesticide Formulation and Sonochemistry. Applied Chemistry courses are normally tailored to the needs of the chemical industry. In fact, the role of chemistry and chemical products in every branch of engineering is expanding greatly. Nowadays, various products of chemical industries are playing an important role in the field of engineering, with increasing number of such products, each successive years. Thus, Chemistry gave birth to a new discipline; Applied Chemistry.

Applied Chemistry is considered as one of the core subjects for students in engineering and technology for developing scientific temper and appreciation of chemical properties of materials, which they have to handle in their professional career.

There is an apparent dearth of exclusive and comprehensive books on Applied Chemistry, as an independent discipline. Hence, this modest effort. Hopefully, this endeavour of ours, would be welcome by all concerned circles and readers would also respond in a healthy critical manner.

— Editor

1

Chemistry's Application in Engineering

Process of Catalycity

Catalycity is a macroscopic parameter to describe surface catalysis (heterogeneous catalysis). It is usually called a material property, due to its strong dependence on the surface material. This naming is not fully justified, since the catalycity not only dependent on the material but also on its environment.

Catalysis

Gases like Nitrogen, Oxygen or mixtures of the like Air consists at low temperatures of molecules. At elevated temperatures they dissociate, i.e. they break apart into single atoms.

$$X_2 \longrightarrow 2X$$

Atoms tends to recombine in the gas phase.

$$2X \longrightarrow X_2$$

This process can be accelerated with the help of a catalyst. The catalyst provides an alternative reaction path, which is in general faster than the original. Such an reaction path could be:

adsorption:

$$X(g) + (s) \longrightarrow X(s)$$

reaction at the surface:

$$2X(s) \longrightarrow X_2(s) + (s)$$

desorption:

$$X_2(s) \longrightarrow X(g) + (s)$$

Definition

Catalycity relates the number of reacting atoms to the number of atoms that hit a surface. Therefore, a catalycity of one means that all impinging particles do recombine.

$$\gamma = \frac{M_{rec}}{M^{\downarrow}}$$

Usage

The parameter catalycity is mainly used in hypersonics, i.e. the science high speed and high temperature flow. These kind of flows occur for instance in Atmospheric re-entry.

Limitations

In mixtures of gases where interactions between the different species occurs, the parameter catalycity is not appropriate.

Determination

The determination of catalycity is usually done in Plasma (physics) wind tunnels like arc-jets or inductively coupled plasma generators.

Reforming of Catalysts

Catalytic reforming is a chemical process used to convert petroleum refinery naphthas, typically having low octane ratings, into high-octane liquid products called reformates which are components of high-octane gasoline (also known as petrol). Basically, the process rearranges or restructures the hydrocarbon molecules in the naphtha feedstocks as well as breaking some of the molecules into smaller molecules.

The overall effect is that the product reformate contains hydrocarbons with more complex molecular shapes having higher octane values than the hydrocarbons in the naphtha feedstock. In so doing, the process separates hydrogen atoms from the hydrocarbon molecules and produces very significant amounts of by-product hydrogen gas for use in a number of the other processes involved in a modern petroleum refinery. Other by-products are small amounts of methane, ethane, propane and butane.

This process is quite different from and not to be confused with the catalytic steam reforming process used industrially to produce various products such as hydrogen, ammonia and methanol from natural gas, naphtha or other petroleum-derived feedstocks. Nor is this process to be confused with various other catalytic reforming processes that use methanol or biomass-derived feedstocks to produce hydrogen for fuel cells or other uses.

Universal Oil Products (also known as UOP) is a multinational company developing and delivering technology to the petroleum refining, natural gas processing, petrochemical production and other manufacturing industries. In the 1940s, an eminent research chemist named Vladimir Haensel working for UOP developed a catalytic reforming process using a catalyst containing platinum. Haensel's process was subsequently commercialised by UOP in 1949 for producing a high octane gasoline from low octane naphthas and the UOP process become known as the Platforming process. The first Platforming unit was built in 1949 at the refinery of the Old Dutch Refining Company in Muskegon, Michigan.

In the years since then, many other versions of the process have been developed by some of the major oil companies and other organisations. Today, the large majority of gasoline produced worldwide is derived from the catalytic reforming process.

To name a few of the other catalytic reforming versions that were developed, all of which utilised a platinum and/or a rhenium catalyst:

- *Rheniforming:* Developed by Chevron Oil Company.
- *Powerforming:* Developed by Esso Oil Company, now known as ExxonMobil.
- *Magnaforming:* Developed by Englehard Catalyst Company and Atlantic Richfield Oil Company.
- *Ultraforming:* Developed by Standard Oil of Indiana, now a part of the British Petroleum Company.
- *Houdriforming:* Developed by the Houdry Process Corporation.
- *CCR Platforming:* A Platforming version, designed for continuous catalyst regeneration, developed by UOP.
- *Octanising:* A catalytic reforming version developed by Axens, a subsidiary of Institut francais du petrole (IFP), designed for continuous catalyst regeneration.

Typical Naphtha Feedstocks

A petroleum refinery includes many unit operations and unit processes. The first unit operation in a refinery is the continuous distillation of the petroleum crude oil being refined. The overhead liquid distillate is called naphtha and will become a major component of the refinery's gasoline (petrol) product after it is further processed through a catalytic hydrodesulphuriser to remove sulphur-containing hydrocarbons and a catalytic reformer to reform its hydrocarbon molecules into more complex molecules with a higher octane rating value. The naphtha is a mixture of very many different hydrocarbon compounds. It has an initial boiling point of about 35°C and a final boiling point of about 200°C, and it contains paraffin, naphthene (cyclic paraffins) and aromatic hydrocarbons ranging from those containing 4 carbon atoms to those containing about 10 or 11 carbon atoms.

The naphtha from the crude oil distillation is often further distilled to produce a "light" naphtha containing most (but not all) of the hydrocarbons with 6 or less carbon atoms and a "heavy" naphtha containing most (but not all) of the hydrocarbons with more than 6 carbon atoms. The heavy naphtha has an initial boiling point of about 140 to 150°C and a final boiling point of about 190 to 205°C. The naphthas derived from the distillation of crude oils are referred to as "straight-run" naphthas.

It is the straight-run heavy naphtha that is usually processed in a catalytic reformer because the light naphtha has molecules with 6 or less carbon atoms which, when reformed, tend to crack into butane and lower molecular weight hydrocarbons which are not useful as high-octane gasoline blending components. Also, the molecules with 6 carbon atoms tend to form aromatics which is undesirable because governmental environmental regulations in a number of countries limit the amount of aromatics (most particularly benzene) that gasoline may contain.

It should be noted that there are a great many petroleum crude oil sources worldwide and each crude oil has its own unique composition or "assay". Also, not all refineries process the same crude oils and each refinery produces its own straight-run naphthas with their own unique initial and final boiling points. In other words, naphtha is a generic term rather than a specific term.

The table just below lists some fairly typical straight-run heavy naphtha feedstocks, available for catalytic reforming, derived from various crude oils. It can be seen that they differ significantly in their content of paraffins, naphthenes and aromatics:

Typical Heavy Naphtha Feedstocks

Crude oil name ⇒ *Location ⇒*	*Barrow Island* *Australia*	*Mutineer-Exeter* *Australia*	*CPC Blend* *Kazakhstan*	*Draugen* *North Sea*
Initial boiling point,°C	149	140	149	150
Final boiling point,°C	204	190	204	180
Paraffins, liquid volume	46	62	57	38
Naphthenes, liquid volume %	42	32	27	45
Aromatics, liquid volume %	12	6	16	17

Some refinery naphthas include olefinic hydrocarbons, such as naphthas derived from the fluid catalytic cracking and coking processes used in many refineries. Some refineries may also desulphurise and catalytically reform those naphthas. However, for the most part, catalytic reforming is mainly used on the straight-run heavy naphthas, such as those in the above table, derived from the distillation of crude oils.

The Reaction Chemistry

There are a good many chemical reactions that occur in the catalytic reforming process, all of which occur in the presence of a catalyst and a high partial pressure of hydrogen. Depending upon the type or version of catalytic reforming used as well as the desired reaction severity, the reaction conditions range from temperatures of about 495 to 525°C and from pressures of about 5 to 45 atm.

The commonly used catalytic reforming catalysts contain noble metals such as platinum and/or rhenium, which are very susceptible to poisoning by sulphur and nitrogen compounds. Therefore, the naphtha feedstock to a catalytic reformer is always preprocessed in a hydrodesulphurisation unit which removes both the sulphur and the nitrogen compounds.

The four major catalytic reforming reactions are:

1. The dehydrogenation of naphthenes to convert them into aromatics as exemplified in the conversion methylcyclohexane (a naphthene) to toluene (an aromatic), as shown below:

Methycylclohexane ⟶ Toluene + $3H_2$

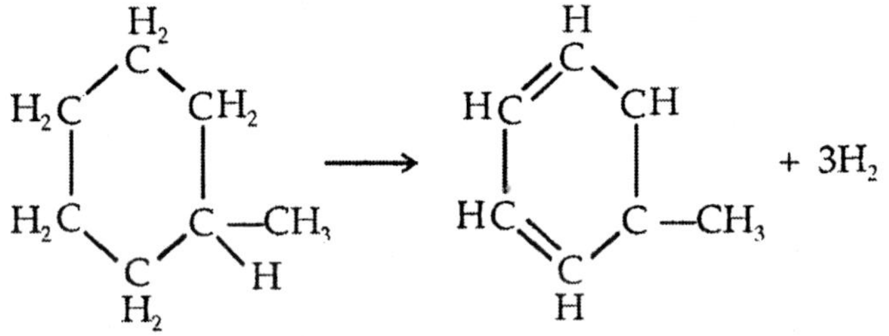

2. The isomerisation of normal paraffins to isoparaffins as exemplified in the conversion of normal octane to 2,5-Dimethylhexane (an isoparaffin), as shown below:

n-Octane ⟶ 2,5-Dimethylhexane

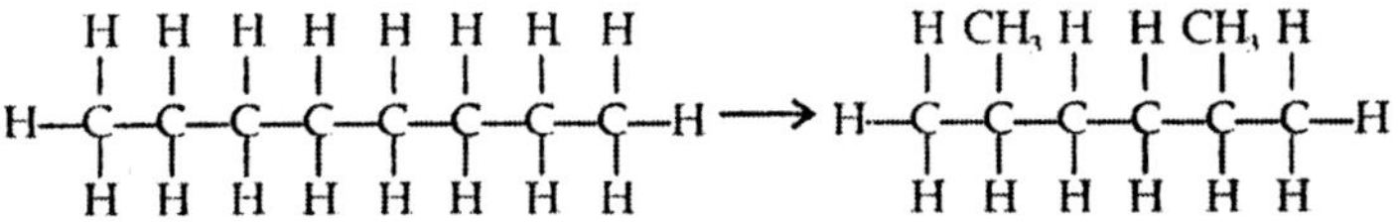

3. The dehydrogenation and aromatisation of paraffins to aromatics (commonly called dehydrocyclisation) as exemplified in the conversion of normal heptane to toluene, as shown below:

n-Heptane ⟶ Toluene + $4H_2$

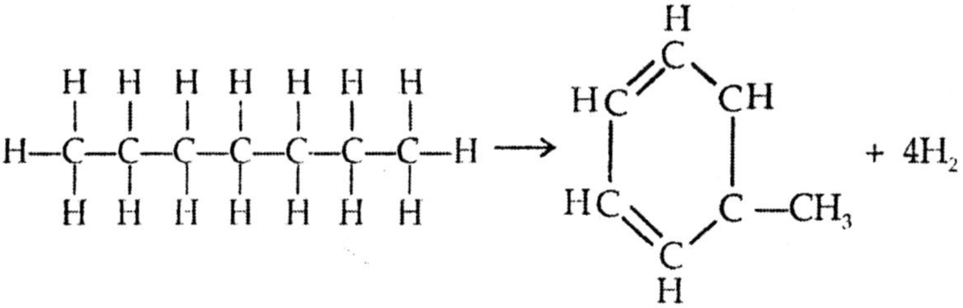

4. The hydrocracking of paraffins into smaller molecules as exemplified by the cracking of normal heptane into isopentane and ethane, as shown below:

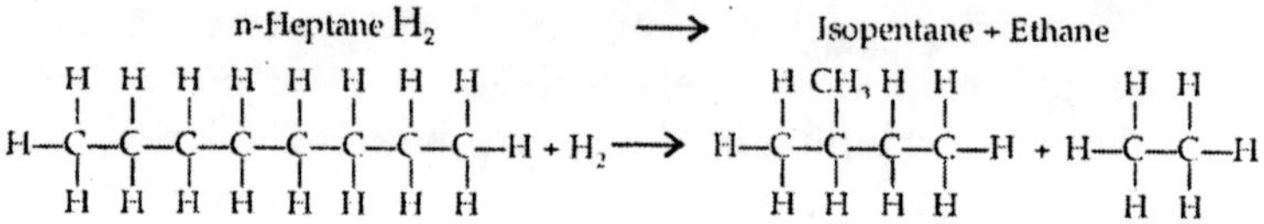

The hydrocracking of paraffins is the only one of the above four major reforming reactions that consumes hydrogen. The isomerisation of normal paraffins does not consume or produce hydrogen. However, both the dehydrogenation of naphthenes and the dehydrocyclisation of paraffins produce hydrogen. The overall net production of hydrogen in the catalytic reforming of petroleum naphthas ranges from about 50 to 200 cubic metres of hydrogen gas (at 0°C and 1 atm) per cubic metre of liquid naphtha feedstock. In the United States customary units, that is equivalent to 300 to 1200 cubic feet of hydrogen gas (at 60°F and 1 atm) per barrel of liquid naphtha feedstock. In many petroleum refineries, the net hydrogen produced in catalytic reforming supplies a significant part of the hydrogen used elsewhere in the refinery (for example, in hydrodesulphurisation processes).

Process Description

The most commonly used type of catalytic reforming unit has three reactors, each with a fixed bed of catalyst, and all of the catalyst is regenerated *in situ* during routine catalyst regeneration shutdowns which occur approximately once each 6 to 24 months. Such units are referred to as semi-regenerative catalytic reformers.

Some catalytic reforming units have an extra spare or swing reactor and each reactor can be individually isolated so that any one reactor can be undergoing in situ regeneration while the other reactors are in operation. When that reactor is regenerated, it replaces another reactor which, in turn, is isolated so that it can then be regenerated. Such units, referred to as cyclic catalytic reformers, are not very common. Cyclic catalytic reformers serve to extend the period between required shutdowns.

The latest and most modern type of catalytic reformers are called *continuous catalyst regeneration (CCR)* reforming units. CCR units are characterised by continuous in situ regeneration of part of the catalyst in a special regenerator, and by continuous addition of the regenerated catalyst to the operating reactors. As of 2006, two CCR versions available: UOP's CCR Platformer process and Axen's Octanising process. The installation and use of CCR units is rapidly increasing.

Many of the earliest catalytic reforming units (in the 1950's and 1960's) were non-regenerative in that they did not perform *in situ* catalyst regeneration. Instead, when needed, the aged catalyst was replaced by fresh catalyst and the aged catalyst was shipped to catalyst manufacturer's to be either regenerated or to recover the platinum content of the aged catalyst. Very few, if any, catalytic reformers currently in operation are non-regenerative.

The liquid feed is pumped up to the reaction pressure (5 to 45 atm) and is joined by a stream of hydrogen-rich recycle gas. The resulting liquid-gas mixture is preheated by flowing through a heat exchanger. The preheated feed mixture is then totally vaporised and heated to the reaction temperature (495 to 520°C) before the vaporised reactants enter the first reactor. As the vaporised reactants flow through the fixed bed of catalyst in the reactor, the major reaction is the dehydrogenation of naphthenes to aromatics which is highly endothermic and results in a large temperature decrease between the inlet and outlet of the reactor. To maintain the required reaction temperature and the rate of reaction, the vaporised stream is reheated in the second fired heater before it flows through the second reactor. The temperature again decreases across the second reactor and the vaporised stream must again be reheated in the third fired heater before it flows through the third reactor. As the vaporised stream proceeds through the three reactors, the reaction rates decrease and the reactors therefore become larger. At the same time, the amount of reheat required between the reactors becomes smaller. Usually, three reactors are all that is required to provide the desired performance of the catalytic reforming unit.

The hot reaction products from the third reactor are partially cooled by flowing through the heat exchanger where the feed to the first reactor is preheated and then flow through a water-cooled heat exchanger before flowing through the pressure controller (PC) into the gas separator.

Most of the hydrogen-rich gas from the gas separator vessel returns to the suction of the recycle hydrogen gas compressor and the net production of hydrogen-rich gas from the reforming reactions is exported for use in the other refinery processes that consume hydrogen (such as hydrodesulphurisation units and/or a hydrocracker unit).

The liquid from the gas separator vessel is routed into a fractionating column commonly called a *stabiliser*. The overhead off-gas product from the stabiliser contains the by-product methane, ethane, propane and butane gases produced by the hydrocracking reactions and it may also contain some small amount of hydrogen. That off-gas is routed to the refinery's central gas processing plant for removal and recovery of propane and butane. The residual gas after such processing becomes part of the refinery's fuel gas system.

The bottoms product from the stabiliser is the high-octane liquid reformate that will become a component of the refinery's product gasoline.

Catalysts and Mechanisms

Most catalytic reforming catalysts contain platinum or rhenium on a silica or silica-alumina support base, and some contain both platinum and rhenium. Fresh catalyst is chlorided (chlorinated) prior to use.

The noble metals (platinum and rhenium) are considered to be catalytic sites for the dehydrogenation reactions and the chlorinated alumina provides the acid sites needed for isomerisation, cyclisation and hydrocracking reactions.

The activity (i.e., effectiveness) of the catalyst in a semi-regenerative catalytic reformer is reduced over time during operation by carbonaceous coke deposition and chloride loss. The activity of the catalyst can be periodically regenerated or restored by *in situ* high temperature oxidation of the coke followed by chlorination. As stated earlier herein, semi-regenerative catalytic reformers are regenerated about once per 6 to 24 months.

Normally, the catalyst can be regenerated perhaps 3 or 4 times before it must be returned to the manufacturer for reclamation of the valuable platinum and/or rhenium content.

Uses of Cement Kiln

Cement kilns are used for the pyroprocessing stage of manufacture of Portland and other types of hydraulic cement. Over a billion tons of cement are made per annum, and the cement kiln is the heart of this production process.

The Manufacture of Cement Clinker

A typical process of manufacture consists of three stages:

- Grinding a mixture of limestone and clay or shale to make a fine "rawmix",
- Heating the rawmix to sintering temperature in a cement kiln, and
- Grinding the resulting clinker to make cement.

In the second stage, the rawmix is fed into the kiln and gradually heated by contact with the hot gases from combustion of the kiln fuel. Successive chemical reactions take place as the temperature of the rawmix rises:

- *70-110°C:* Free water is evaporated.
- *400-600°C:* clay-like minerals are decomposed into their constituent oxides; principally SiO_2 and Al_2O_3. Dolomite ($CaMg(CO_3)_2$) decomposes to calcium carbonate, MgO and CO_2.
- *650-900°C:* Calcium carbonate reacts with SiO_2 to form belite (Ca_2SiO_4).
- *900-1050°C:* The remaining calcium carbonate decomposes to calcium oxide and CO_2.
- *1300-1450°C:* Partial (20-30 per cent) melting takes place, and belite reacts with calcium oxide to form alite ($Ca_3O.SiO_4$).

Alite is the characteristic constituent of Portland cement. Typically, a peak temperature of 1400-1450°C is required to complete the reaction. The partial melting causes the material to aggregate into lumps or nodules, typically of diameter 1-10 mm. This is called clinker. The hot clinker next falls into a cooler which recovers most of its heat, and cools the clinker to around 100°C, at which temperature it can be conveniently conveyed to storage. The cement kiln system is designed to accomplish these processes efficiently.

Early History

Portland cement clinker was first made (in 1842) in a modified form of the traditional static lime kiln. The basic, egg-cup shaped lime kiln was provided with a conical or beehive shaped extension to increase draught and thus obtain the higher temperature needed to make cement clinker. For nearly half a century, this design, and minor modifications, remained the only method of manufacture. The kiln was restricted in size by the strength of the chunks of rawmix: if the charge in the kiln collapsed under its own weight, the kiln would be extinguished. For this reason, beehive kilns never made more than 30 tons of clinker per batch. A batch took one week to turn around: a day to fill the kiln, three days to burn off, two days to cool, and a day to unload. Thus, a kiln would produce about 1500 tons per year.

Around 1885, experiments began on design of continuous kilns. One design was the shaft kiln, similar in design to a blast furnace. Rawmix in the form of lumps and fuel were continuously added at the top, and clinker was continually withdrawn at the bottom.

Air was blown through under pressure from the base to combust the fuel. The shaft kiln had a brief period of use before it was eclipsed by the rotary kiln, but it had a limited renaissance from 1970 onward in China and elsewhere, when it was used for small-scale, low-tech plants in rural areas away from transport routes. Several thousand such kilns were constructed in China. A typical shaft kiln produces 100-200 tons per day.

From 1885, trials began on the development of the rotary kiln, which today accounts for more than 95 per cent of world production.

The Rotary Kiln

The rotary kiln consists of a tube made from steel plate, and lined with firebrick. The tube slopes slightly (1-4°) and slowly rotates on its axis at between 30 and 250 revolutions per hour. Rawmix is fed in at the upper end, and the rotation of the kiln causes it to gradually move downhill to the other end of the kiln. At the other end fuel, in the form of gas, oil, or pulverised solid fuel, is blown in through the "burner pipe", producing a large concentric flame in the lower part of the kiln tube. As material moves under the flame, it reaches its peak temperature, before dropping out of the kiln tube into the cooler. Air is drawn first through the cooler and then through the kiln for combustion of the fuel. In the cooler the air is heated by the cooling clinker, so that it may be 400-800°C before it enters the kiln, thus causing intense and rapid combustion of the fuel.

The earliest successful rotary kilns were developed in Pennsylvania around 1890, and were about 1.5 m in diameter and 15 m in length. Such a kiln made about 20 tons of clinker per day. The fuel, initially, was oil, which was readily available in Pennsylvania at the time. It was particularly easy to get a good flame with this fuel. Within the next 10 years, the technique of firing by blowing in pulverised coal was developed, allowing the use of the cheapest available fuel. By 1905, the largest kilns were 2.7 x 60 m in size, and made 190 tons per day. At that date, after only 15 years of development, rotary kilns accounted for half of world production. Since then, the capacity of kilns has increased steadily, and the largest kilns today produce around 10,000 tons per day. In contrast to static kilns, the material passes through quickly: it takes from 3 hours (in some old wet process kilns) to as little as 10 minutes (in short precalciner kilns). Rotary kilns run 24 hours a day, and are typically stopped only for a few days once or twice a year for essential maintenance. This is an important discipline, because heating up and cooling down are long, wasteful and damaging processes. Uninterrupted runs as long as 18 months have been achieved.

The Wet Process and the Dry Process

From the earliest times, two different methods of rawmix preparation were used: the mineral components were either dry-ground to form a flour-like powder, or were wet-ground with added water to produce a fine slurry with the consistency of paint, and with a typical water content of 40-45 per cent.

The wet process suffered the obvious disadvantage that, when the slurry was introduced into the kiln, a large amount of extra fuel was used in evaporating the water. Furthermore, a larger kiln was needed for a given clinker output, because much of the kiln's length was used up for the drying process. On the other hand, the wet process had a number of advantages. Wet grinding of hard minerals is usually much more efficient than dry grinding. When slurry is dried in the kiln, it forms a granular crumble that is ideal for subsequent heating in the kiln. On the other hand, in the dry process, it is very difficult to keep the fine powder rawmix in the kiln, because the fast-flowing combustion gases tend to blow it back out again.

It became a practice to spray water into dry kilns in order to "damp down" the dry mix, and thus, for many years there was little difference in efficiency between the two processes, and the overwhelming majority of kilns used the wet process. By 1950, a typical large, wet process kiln, fitted with drying-zone heat exchangers, was 3.3 x 120 m in size, made 680 tons per day, and used about 0.25-0.30 tons of coal fuel for every tonne of clinker produced. Before the energy crisis of the 1970s put an end to new wet-process installations, kilns as large as 5.8 x 225 m in size were making 3000 tons per day.

An interesting footnote on the wet process history is that some manufacturers have in fact made very old wet process facilities profitable through the use of waste fuels. Plants that burn waste fuels enjoy a negative fuel cost (they are paid by industries needing to dispose of materials that have energy content and can be safely disposed of in the cement kiln thanks to its high temperatures and longer retention times). As a result the inefficiency of the wet process is an advantage—to the manufacturer.

By locating waste burning operations at older wet process locations, higher fuel consumption actually equates to higher profits for the manufacturer, although it produces correspondingly greater environmental damage. Environmentally concerned manufacturers are all abandoning the use of wet process in response to the need to minimise CO_2 emissions.

Preheaters

In the 1930s, significantly, in Germany, the first attempts were made to redesign the kiln system to minimise waste of fuel. This led to two significant developments:

- The grate preheater.
- The gas-suspension preheater.

Grate Preheaters

The grate preheater consists of a chamber containing a chain-like high-temperature steel moving grate, attached to the cold end of the rotary kiln. A dry-powder rawmix is turned into a hard pellets of 10-20 mm diameter in a nodulising pan, with the addition of 10-15 per cent water. The pellets are loaded onto the moving grate, and the hot combustion gases from the rear of the kiln are passed through the bed of pellets from beneath. This dries and partially calcines the rawmix very efficiently.

The pellets then drop into the kiln. Very little powdery material is blow out of the kiln. Because the rawmix is damped in order to make pellets, this is referred to as a "semi-dry" process. The grate preheater is also applicable to the "semi-wet" process, in which the rawmix is made as a slurry, which is first de-watered with a high-pressure filter, and the resulting "filter-cake" is extruded into pellets, which are fed to the grate. In this case, the water content of the pellets is 17-20 per cent. Grate preheaters were most popular in the 1950s and 60s, when a typical system would have a grate 28 m long and 4 m wide, and a rotary kiln of 3.9 x 60 m, making 1050 tons per day, using about 0.11-0.13 tons of coal fuel for every tonne of clinker produced. Systems up to 3000 tons per day were installed.

Gas-suspension Preheaters

The key component of the gas-suspension preheater is the cyclone. A cyclone is a conical vessel into which a dust-bearing gas-stream is passed tangentially. This produces a vortex within the vessel. The gas leaves the vessel through a coaxial "vortex-finder". The solids are thrown to the outside edge of the vessel by centrifugal action, and leave through a valve in the vertex of the cone. Cyclones were originally used to clean up the dust-laden gases leaving simple dry process kilns.

If, instead, the entire feed of rawmix is encouraged to pass through the cyclone, it is found that a very efficient heat exchange takes place: the gas is efficiently cooled, hence producing less waste of heat to the atmosphere, and the rawmix is efficiently heated. This efficiency is further increased if a number of cyclones are connected in series.

The number of cyclones stages used in practice varies from 1 to 6. Energy, in the form of fan-power, is required to draw the gases through the string of cyclones, and at a string of 6 cyclones, the cost of the added fan-power needed for an extra cyclone exceeds the efficiency advantage gained. It is normal to use the warm exhaust gas to dry the raw materials in the rawmix grinding mill, and if the raw materials are wet, hot gas from a less efficient preheater is desirable.

For this reason, the most commonly encountered suspension preheaters have 4 cyclones. The hot feed that leaves the base of the preheater string is typically 20 per cent calcined, so the kiln has less subsequent processing to do, and can therefore achieve a higher specific output. Typical large systems installed in the early 1970s had cyclones 6 m in diameter, a rotary kiln of 5 x 75 m, making 2500 tons per day, using about 0.11-0.12 tons of coal fuel for every tonne of clinker produced.

Development of Precalciners

In the 1970s the precalciner was pioneered in Japan, and has subsequently become the equipment of choice for new large installations worldwide. The precalciner is a development of the suspension preheater. The philosophy is this: the amount of fuel that can be burned in the kiln is directly related to the size of the kiln. If part of the fuel necessary to burn the rawmix is burned outside the kiln, the output of the system can be increased for a given kiln size. Users of suspension preheaters found that output could be increased by injecting extra fuel into the base of the preheater.

The logical development was to install a specially designed combustion chamber at the base of the preheater, into which pulverised coal is injected. This is referred to as an "air-through" precalciner, because the combustion air for both the kiln fuel and the calciner fuel all passes through the kiln. This kind of precalciner can burn up to 30 per cent (typically 20 per cent) of its fuel in the calciner. If more fuel were injected in the calciner, the extra amount of air drawn through the kiln would cool the kiln flame excessively. The feed is 40-60 per cent calcined before it enters the rotary kiln.

The ultimate development is the "air-separate" precalciner, in which the hot combustion air for the calciner arrives in a duct directly from the cooler, bypassing the kiln. Typically, 60-75 per cent of the fuel is burned in the precalciner. In these systems, the feed entering the rotary kiln is 100 per cent calcined. The kiln has only to raise the feed to sintering temperature. In theory the maximum efficiency would be achieved if all the fuel were burned in the preheater, but the sintering operation involves partial melting and nodulisation to make clinker, and the rolling action of the rotary kiln remains the most efficient way of doing this. Large modern installations typically have two parallel strings of 4 or 5 cyclones, with one attached to the kiln and the other attached to the precalciner chamber. A rotary kiln of 6 × 100 m makes 8-10,000 tons per day, using about 0.10-0.11 tons of coal fuel for every tonne of clinker produced. The kiln is dwarfed by the massive preheater tower and cooler in these installations.

Such a kiln produces 3 million tons of clinker per year, and consumes 300,000 tons of coal. 6 m diameter appears to be the limit of size of rotary kilns, because the flexibility of the steel shell becomes unmanageable at or above this size, and the firebrick lining tends to fail when the kiln flexes.

Ancillary Equipment

Essential equipment in addition to the kiln tube and the preheater are:

- Cooler,
- Fuel Mills,
- Fans, and
- Exhaust Gas Cleaning Equipment.

Coolers

Early systems used rotary coolers, which were rotating cylinders similar to the kiln, into which the hot clinker dropped. The combustion air was drawn up through the cooler as the clinker moved down, cascading through the air stream. In the 1920s, satellite coolers became common and remained in use until recently. These consist of a set (typically 7-9) of tubes attached to the kiln tube. They have the advantage that they are sealed to the kiln, and require no separate drive. From about 1930, the grate cooler was developed.

This consists of a perforated grate through which cold air is blown, enclosed in a rectangular chamber. A bed of clinker up to 0.5 m deep moves along the grate. These coolers have two main advantages: they cool the clinker rapidly, which is desirable from a quality point of view, and, because they don't rotate, hot air can be ducted out of them for use in fuel drying, or for use as precalciner combustion air. The latter advantage means that they have become the only type used in modern systems.

Fuel Mills

Fuel systems are divided into two categories:

- Direct firing
- Indirect firing

In direct firing, the fuel is fed at a controlled rate to the fuel mill, and the fine product is immediately blown into the kiln. The advantage of this system is that it is not necessary to store the hazardous ground fuel: it is used as soon as it is made. For this reason it was the system of choice for older kilns. A disadvantage is that the fuel mill has to run all the time: if it breaks down, the kiln has to stop if no backup system is available.

In indirect firing, the fuel is ground by an intermittently-run mill, and the fine product is stored in a silo of sufficient size to supply the kiln though fuel mill stoppage periods. The fine fuel is metred out of the silo at a controlled rate and blown into the kiln. This method is now favoured for precalciner systems, because both the kiln and the precalciner can be fed with fuel from the same system. Special techniques are required to store the fine fuel safely, and coals with high volatiles are normally milled in an inert atmosphere (e.g. CO_2).

Fans

A large volume of gases has to be moved through the kiln system. Particularly in suspension preheater systems a high degree of suction has to be developed at the exit of the system to drive this. Fans are also used to force air through the cooler bed, and to propel the fuel into the kiln. Fans account for most of the electric power consumed in the system, typically amounting to 10-15 kWh per tonne of clinker.

Gas Cleaning

The exhaust gases from a modern kiln typically amount to 2 tons (or 1500 cubic metres at NTP) per tonne of clinker made. The gases carry a large amount of dust—typically

30 g per cubic metre. Environmental regulations specific to different countries require that this be reduced to (typically) 0.1 g per cubic metre, so dust capture needs to be at least 99.7 per cent efficient. Methods of capture include electrostatic precipitators and bag-filters.

Kiln Fuels

Fuels that have been used for primary firing include coal, petroleum coke, heavy fusel oil, natural gas, landfill off-gas and oil refinery flare gas. High carbon fuels such as coal are preferred for kiln firing, because they yield a luminous flame. The clinker is brought to its peak temperature mainly by radiant heat transfer, and a bright (i.e. high emissivity) and hot flame is essential for this. In favourable circumstances, high-rank bituminous coal can produce a flame at 2050°C. Natural gas can only produce a flame of, at best 1950°C, and this is also less luminous, so it tends to result in lower kiln output.

In addition to these primary fuels, various combustible waste materials have been fed to kilns, notably used tyres, which are very difficult to dispose of by other means. In theory, cement kilns are an attractive way of disposing of hazardous materials, because of:

- The temperatures in the kiln, which are much higher than in other combustion systems (e.g. incinerators),
- The alkaline conditions in the kiln, afforded by the high-calcium rawmix, which can absorb acidic combustion products,
- The ability of the clinker to absorb heavy metals into its structure.

Whole tyres are commonly introduced in the kiln, by rolling them into the upper end of a preheater kiln, or by dropping them through a slot midway along a long wet kiln. In either case, the high gas temperatures (1000-1200°C) cause almost instantaneous, complete and smokeless combustion of the tyre. Alternatively, tyres are chopped into 5-10 mm chips, in which form they can be injected into a precalciner combustion chamber. The steel and zinc in the tyres become chemically incorporated into the clinker. Other wastes have included solvents and clinical wastes. A very high level of monitoring of both the fuel and its combustion products is necessary to maintain safe operation.

For maximum kiln efficiency, high quality conventional fuels are the best choice. When using waste materials, in order to avoid prohibited emissions (e.g. of dioxins) it is necessary to control the kiln system in a manner that is non-optimal for efficiency and output, and coarse combustibles such as tyres can cause major product quality problems.

Kiln Control

The objective of kiln operation is to make clinker with the required chemical and physical properties, at the maximum rate that the size of kiln will allow, while meeting environmental standards, at the lowest possible operating cost. The kiln is very sensitive to control strategies, and a poorly run kiln can easily double cement plant operating costs

Formation of the desired clinker minerals involves heating the rawmix through the temperature stages. The finishing reaction that takes place in the hottest part of the kiln, under the flame, is:

$$Ca_2SiO_4 + CaO \longrightarrow Ca_3SiO_5$$

If the reaction is incomplete, excessive amounts of free calcium oxide remain in the clinker. Regular measurement of the free CaO content is used as a means of tracking the clinker quality. As a parameter in kiln control, free CaO data is somewhat ineffective because, even with fast automated sampling and analysis, the data, when it arrives, may be 10 minutes "out of date", and more immediate data must be used for minute-to-minute control.

Conversion of belite to alite requires partial melting, the resulting liquid being the solvent in which the reaction takes place. The amount of liquid, and hence the speed of the finishing reaction, is related to temperature. To meet the clinker quality objective, the most obvious control is that the clinker should reach a peak temperature such that the finishing reaction takes place to the required degree.

A further reason to maintain constant liquid formation in the hot end of the kiln is that the sintering material forms a dam that prevents the cooler upstream feed from flooding out of the kiln. The feed in the calcining zone, because it is a powder evolving carbon dioxide, is extremely fluid.

Cooling of the burning zone, and loss of unburned material into the cooler, is called "flushing", and in addition to causing lost production can cause massive damage.

However, for efficient operation, steady conditions need to be maintained throughout the whole kiln system. The feed at each stage must be at a temperature such that it is "ready" for processing in the next stage. To ensure this, the temperature of both feed and gas must be optimised and maintained at every point. The external controls available to achieve this are few:

- *Feed Rate:* This defines the kiln output.
- *Rotary Kiln Speed:* This controls the rate at which the feed moves through the kiln tube.
- *Fuel Injection Rate:* This controls the rate at which the "hot end" of the system is heated.
- *Exhaust fan Speed or Power:* This controls gas flow, and the rate at which heat is drawn from the "hot end" of the system to the "cold end".

In the case of precalciner kilns, further controls are available:

- Independent control of fuel to kiln and calciner.
- Independent fan controls where there are multiple preheater strings.

The independent use of fan speed and fuel rate is constrained by the fact that there must always be sufficient oxygen available to burn the fuel, and in particular, to burn carbon to carbon dioxide. If carbon monoxide is formed, this represents a waste of fuel, and also indicates reducing conditions within the kiln which must be avoided at all costs since it causes destruction of the clinker mineral structure. For this reason, the exhaust gas is continually analysed for O_2, CO, NO and SO_2.

The assessment of the clinker peak temperature has always been problematic. Contact temperature measurement is impossible because of the chemically aggressive and abrasive nature of the hot clinker, and optical methods such as infrared pyrometry are difficult because of the dust and fume-laden atmosphere in the burning zone.

The traditional method of assessment was to view the bed of clinker and deduce the amount of liquid formation by experience. As more liquid forms, the clinker becomes stickier, and the bed of material climbs higher up the rising side of the kiln. It is usually also possible to assess the length of the zone of liquid formation, beyond which powdery "fresh" feed can be seen. Cameras, with or without infrared measurement capability, are mounted on the kiln hood to facilitate this.

On many kilns, the same information can be inferred from the kiln motor power drawn, since sticky feed riding high on the kiln wall increases the eccentric turning load of the kiln. The formation of NO from nitrogen and oxygen takes place only at high temperatures, and so the NO level gives an indication of the combined feed and flame temperature. SO_2 is formed by thermal decomposition of calcium sulphate in the clinker, and so also gives in indication of clinker temperature. Modern computer control systems usually make a "calculated" temperature, using contributions from all these information sources, and then set about controlling it.

As an exercise in process control, kiln control is extremely challenging, because of multiple interrelated variables, non-linear responses, and variable process lags. Computer control systems were first tried in the early 1960s, initially with poor results due mainly to poor process measurements. Since 1990, complex high level supervisory control systems have been standard on new installations. These operate using expert system strategies, that maintain a "just sufficient" burning zone temperature, below which the kiln's operating condition will deteriorate catastrophically, thus requiring rapid-response, "knife-edge" control.

Treatment of Amine Gas

Amine gas treating refers to a group of processes that use aqueous solutions of various amines to remove hydrogen sulphide (H_2S) and carbon dioxide (CO_2) from gases. It is a common unit process used in refineries, petrochemical plants, natural gas processing plants and other industries. The process is also known as Acid gas removal and Gas sweetening.

Processes within oil refineries or natural gas processing plants that remove hydrogen sulphide and/or mercaptans are commonly referred to as *sweetening* processes because they results in products which no longer have the sour, foul odours of mercaptans and hydrogen sulphide.

There are many different amines used in gas treating:

- Monoethanolamine (MEA)
- Diethanolamine (DEA)
- Methyldiethanolamine (MDEA)
- Di-isopropylamine (DIPA)
- Diglycolamine (DGA)

The most commonly used amines in industrial plants are the alkanolamines MEA and DEA.

Amines are also used in many oil refineries to remove sour gases from liquid hydrocarbons such as liquified petroleum gas (LPG).

Description of a Typical Amine Treater

Gases containing H_2S or both H_2S and CO_2 are commonly referred to as *sour gases* or *acid gases* in the hydrocarbon processing industries.

A typical amine gas treating process includes an absorber unit and a regenerator unit as well as accessory equipment. In the absorber, the downflowing amine solution absorbs H_2S and CO_2 from the upflowing sour gas to produce a sweetened gas stream (i.e., an H_2S-free gas) as a product and an amine solution rich in the absorbed acid gases.

The resultant "rich" amine is then routed into the regenerator (a stripper with a reboiler) to produce regenerated or "lean" amine that is recycled for reuse in the absorber. The stripped overhead gas from the regenerator is concentrated H_2S and CO_2.

In oil refineries, that stripped gas is mostly H_2S, much of which often comes from a sulphur-removing process called hydrodesulphurisation. This H_2S-rich stripped gas stream is then usually routed into a Claus process to convert it into elemental sulphur. In fact, the vast majority of the 64,000,000 metric tons of sulphur produced worldwide in 2005 was by-product sulphur from refineries and other hydrocarbon processing plants. In some plants, more than one amine absorber unit may share a common regenerator unit.

In the steam reforming process of hydrocarbons to produce gaseous hydrogen for subsequent use in the industrial synthesis of ammonia, amine treating is one of the commonly used processes for removing excess carbon dioxide in the final purification of the gaseous hydrogen.

Storage and Capture of Carbon

Carbon capture and storage (CCS) is an approach to mitigating climate change by capturing carbon dioxide (CO_2) from large point sources such as power plants and subsequently storing it instead of releasing it into the atmosphere. Technology for capturing of CO_2 is already commercially available for large CO_2 emitters, such as power plants; however, capture is meaningless without storage.

Storage of CO_2, on the other hand, is a relatively untried concept and as yet (2007) no power plant operates with a full carbon capture and storage system. Currently, the United States government has approved the construction of the world's first CCS power plant, FutureGen.

BP had announced the development of a 350 MW carbon capture and storage plant in Scotland, in which the carbon from a natural gas fired generator plant would be stripped out and pumped into the Miller field in the North Sea. BP have since abandoned this project stating that government plans to support such developments would occur too late in the schedule.

CCS applied to a modern conventional power plant could reduce CO_2 emissions to the atmosphere by approximately 80-90 per cent compared to a plant without CCS. Capturing and compressing CO_2 requires much energy and would increase the fuel needs of a plant with CCS by about 10-40 per cent. These and other system costs are estimated to increase the cost of energy from a power plant with CCS by 30-60 per cent depending on the specific circumstances.

Storage of the CO_2 is envisaged either in deep geological formations, deep oceans, or in the form of mineral carbonates. In the case of deep ocean storage, there is a risk of greatly increasing the problem of ocean acidification, a problem that also stems from the excess of carbon dioxide already in the atmosphere and oceans.

Geological formations are currently considered the most promising sequestration sites, and these are estimated to have a storage capacity of at least 2000 Gt CO_2. IPCC estimates that the economic potential of CCS could be between 10 per cent and 55 per cent of the total carbon mitigation effort until year 2100.

Cost of CCS

Capturing and compressing CO_2 requires much energy, significantly raising the running costs of CCS-equipped power plants. In addition there are added investment or capital costs. The process would increase the energy needs of a plant with CCS by about 10-40 per cent. The costs of storage and other system costs are estimated to increase the costs of energy from a power plant with CCS by 30-60 per cent, depending on the specific circumstances.

Costs of Energy with and without CCS (2002 US$ per kWh)

	Natural gas combined cycle	*Pulverised coal gasification*	*Integrated combined cycle*
Without capture (reference plant)	0.03 - 0.05	0.04 - 0.05	0.04 - 0.06
With capture and geological storage	0.04 - 0.08	0.06 - 0.10	0.06 - 0.09
With capture and Enhanced oil recovery	0.04 - 0.07	0.05 - 0.08	0.04 - 0.08

All costs refer to costs for energy from newly built, large-scale plants. Natural gas combined cycle costs are based on natural gas prices of 2.80-4.40 US$ per GJ (LHV based). Energy costs for PC and IGCC are based on bituminous coal costs of 1.00-1.50 US$ per GJ (LHV. Note that the costs are very dependent on fuel prices (which change continuously), in addition to other factors such as capital costs. Also note that for EOR, the savings are greater for higher oil prices.

The cost of CCS depends on the cost of capture and storage which vary according to the method used. Geological storage in saline formations or depleted oil or gas fields typically cost 0.5-8 US$ per tonne of CO_2 injected, plus an additional 0.1-0.3 US$ for monitoring costs. However, when storage is combined with enhanced oil recovery to extract extra oil from an oil field, the storage could yield net benefits of 10-16 US$ per tonne of CO_2 injected (based on 2003 oil prices). The benefits do not outweigh the extra costs of capture.

Comparisons of CCS with other energy sources can be found in wind energy, solar energy, and Economics of new nuclear power plants.

Environmental Effects

The merit of CCS systems is the reduction of CO_2 emissions by up to 90 per cent, depending on plant type.

Generally, environmental effects from use of CCS arise during power production, CO_2 capture, transport and storage.

Additional energy is required for CO_2 capture, and this means that substantially more fuel has to be used, depending on the plant type.

For new supercritical pulverised coal (PC) plants using current technology, the extra energy requirements range from 24-40 per cent, while for natural gas combined cycle (NGCC) plants the range is 11-22 per cent and for coal-based gasification combined cycle (IGCC) systems it is 14-25 per cent. Obviously, fuel use and environmental problems arising from mining and extraction of coal or gas increase accordingly.

Plants equipped with flue gas desulphurisation (FGD) systems for SO_2 control require proportionally greater amounts of limestone, and systems equipped with SCR systems for NO_X require proportionally greater amounts of ammonia.

IPCC has provided estimates of air emissions from various CCS plant designs. While CO_2 is drastically reduced (though never completely captured), emissions of air pollutants increase significantly, generally due to the energy penalty of capture. Hence, the use of CCS entails a reduction in air quality.

Emissions to air from plants with or without CCS (kg/MWh)

	Natural gas combined cycle	*Pulverised coal gasification*	*Integrated combined cycle*
CO_2	43 (-89 per cent)	107 (-87%)	97 (-88%)
NO_X	0.11 (+22%)	0.77 (+31%)	0.1 (+11%)
SO_X		0.001 (-99,7%)	0.33 (+17,9%)
Ammonia	0.002 (before: 0)	0.23 (+2200%)	

CO_2 Capture

Capturing CO_2 can be applied to large point sources, such as large fossil fuel or biomass energy facilities, industries with major CO_2 emissions, natural gas processing, synthetic fuel plants and fossil fuel-based hydrogen production plants. Broadly, three different types of technologies exist: Post-combustion, precombustion, and oxyfuel combustion.

In post-combustion, the CO_2 is removed after combustion of the fossil fuel—this is the scheme that would be applied to conventional power plants. Here, carbon dioxide is captured from flue gases at power stations (in the case of coal, this is sometimes known as "clean coal"). The technology is well understood and is currently used in niche markets.

The technology for precombustion is widely applied in fertilizer, chemical, gaseous fuel (H_2, CH_4), and power production. In these cases, the fossil fuel is gasified and the resulting CO_2 can be captured from a relatively pure exhaust stream.

In oxyfuel combustion the lignite is burned in oxygen instead of air. This produces a flue gas consisting only carbon dioxide and water vapour, which is cooled and condensed. The result is an almost pure carbon dioxide stream that can be transported to the sequestration site and stored. The technique is promising, but the initial air separation step demands a lot of energy.

An alternate method, which is under development, is chemical looping combustion (CLC). Chemical looping uses a metal oxide as a solid oxygen carrier. Metal oxide particles react with a solid, liquid or gaseous fuel in a fluidised bed combustor, producing solid metal particles and a mixture of carbon dioxide and water vapour.

The water vapour is condensed, leaving pure carbon dioxide which can be sequestered. The solid metal particles are circulated to another fluidised bed where they react with air, producing heat and regenerating metal oxide particles that are recirculated to the fluidised bed combustor.

A few engineering proposals have been made for the much more difficult task of capturing CO_2 directly from the air, but work in this area is speculative and conceptual at this point. Capture costs are estimated to be much higher than from point sources, but may be feasible for dealing with emissions from diffuse sources like automobiles and aircraft.

CO_2 Transport

After capture, the CO_2 must be transported to suitable storage sites. This is done by pipeline, which is generally the cheapest form of transport, or by ship when no pipelines are available. Both methods are currently used for transporting CO_2 for other applications.

CO_2 Storage

Various forms have been conceived for permanent storage of CO_2. These forms include gaseous storage in various deep geological formations (including saline formations and exhausted gas fields), liquid storage in the ocean, and solid storage by reaction of CO_2 with metal oxides to produce stable carbonates.

Geological Storage

Also known as *geosequestration,* this method involves injecting carbon dioxide directly into underground geological formations. Oil fields, gas fields, saline formations, unminable coal seams, and saline-filled basalt formations have been suggested as storage sites.

Here, various physical (e.g., highly impermeable caprock) and geochemical trapping mechanisms would prevent the CO_2 from escaping to the surface. CO_2 is sometimes injected into declining oil fields to increase oil recovery (enhanced oil recovery). This option is attractive because the storage costs are offset by the sale of additional oil that is recovered. Disadvantages of old oil fields are their geographic distribution and their limited capacity.

Unminable coal seams can be used to store CO_2 because CO_2 adsorbs to the surface of coal. However, the technical feasibility depends on the permeability of the coal bed. In the process of absorption the coal releases previously absorbed methane, and the methane can be recovered (Enhanced Coal Bed Methane recovery). The sale of the methane can be used to offset the cost of the CO_2 storage.

Saline formations contain highly mineralised brines, and have so far been considered of no benefit to humans. Saline aquifers have been used for storage of chemical waste in a few cases. The main advantage of saline aquifers is their large potential storage volume and their common occurrence. This will reduce the distances over which CO_2 has to be transported. The major disadvantage of saline aquifers is that relatively little is known about them, compared to oil fields.

To keep the cost of storage acceptable the geophysical exploration may be limited, resulting in larger uncertainty about the aquifer tructure. Unlike storage in oil fields or coal beds no side product will offset the storage cost. Leakage of CO_2 back into the atmosphere may be a problem in saline aquifer storage. However, current research shows that several *trapping mechanisms* immobilise the CO_2 underground, reducing the risk of leakage.

For well-selected, designed and managed geological storage sites, IPCC estimates that CO_2 could be trapped for millions of years, and the sites are likely to retain over 99 per cent of the injected CO_2 over 1,000 years.

Example CCS Projects

As of 2005, three industrial-scale storage projects are in operation. Sleipner is the oldest project (1996) and is located in the North Sea where Norway's Statoil strips carbon dioxide from natural gas with amine solvents and disposes of this carbon dioxide in a saline formation. The carbon dioxide is a waste product of the field's natural gas production and the gas contains more (9 per cent CO_2) than is allowed into the natural gas distribution network.

Storing it underground avoids this problem and saves Statoil hundreds of millions of euro in avoided carbon taxes. Sleipner stores about one million tons CO_2 a year. The Weyburn project started in 2000 and is located in an oil reservoir discovered in 1954 in Weyburn, Southeastern Saskatchewan, Canada. The CO_2 for this project is captured at the Great Plains Coal Gasification plant in Beulah, North Dakota which has produced methane from coal for more than 30 years. At Weyburn, the CO_2 will also be used for enhanced oil recovery with an injection rate of about 1.5 million tons per year. The third site is In Salah, which like Sleipner is a natural gas reservoir located in In Salah, Algeria. The CO_2 will be separated from the natural gas and re-injected into the subsurface at a rate of about 1.2 million tons per year.

Ocean Storage

Another proposed form of carbon storage is in the oceans. Two main concepts exist. The 'dissolution' type injects CO_2 by ship or pipeline into the water column at depths of 1000 m or more, and the CO_2 subsequently dissolves. The 'lake' type deposits CO_2 directly onto the sea floor at depths greater than 3000 m, where CO_2 is denser than water and is expected to form a 'lake' that would delay dissolution of CO_2 into the environment. A third concept is to convert the CO_2 to bicarbonates (using limestone) or hydrates.

The environmental effects of ocean storage are generally negative, but poorly understood. Large concentrations of CO_2 kills ocean organisms, but another problem is that dissolved CO_2 would eventually equilibrate with the atmosphere, so the storage would not be permanent. Also, as part of the CO_2 reacts with the water to form carbonic acid, H_2CO_3, the acidity of the ocean water increases. The resulting environmental effects on benthic life forms of the bathypelagic, abyssopelagic and hadopelagic zones are poorly understood. Even though life appears to be rather sparse in the deep ocean basins, energy and chemical effects in these deep basins could have far reaching implications. Much more work is needed here to define the extent of the potential problems.

The time it takes water in the deeper oceans to circulate to the surface has been estimated to be on the order of 1600 years, varying upon currents and other changing conditions. Costs for deep ocean disposal of liquid CO_2 are estimated at 40-80USD/ton. This figure covers the cost of sequestration at the powerplant and naval transport to the disposal site.

The bicarbonate approach would reduce the pH effects and enhance the retention of CO_2 in the ocean, but this would also increase the costs and other environmental effects.

An additional method of long term ocean based sequestration is to gather crop residue such as corn stalks or excess hay into large weighted bales of biomass and deposit it in the alluvial fan areas of the deep ocean basin. Dropping these residues in alluvial fans would cause the residues to be quickly buried in silt on the sea floor, sequestering the biomass for very long time spans. Alluvial fans exist in all of the world's oceans and seas where river deltas fall off the edge of the continental shelf such as the Mississippi alluvial fan in the gulf of Mexico and the Nile alluvial fan in the Mediterranean Sea.

Mineral Storage

"Carbon sequestration by reacting naturally occurring Mg and Ca containing minerals with CO_2 to form carbonates has many unique advantages. Most notably is the fact that carbonates have a lower energy state than CO_2, which is why mineral carbonation is thermodynamically favourable and occurs naturally (e.g., the weathering of rock over geologic time periods). Secondly, the raw materials such as magnesium based minerals are abundant. Finally, the produced carbonates are unarguably stable and thus re-release of CO_2 into the atmosphere is not an issue. However, conventional carbonation pathways are slow under ambient temperatures and pressures. The significant challenge being addressed by this effort is to identify an industrially and environmentally viable carbonation route that will allow mineral sequestration to be implemented with acceptable economics."

In this process, CO_2 is exothermically reacted with abundantly available metal oxides which produces stable carbonates. This process occurs naturally over many years and is responsible for much of the surface limestone. The reaction rate can be made faster, for example by reacting at higher temperatures and/or pressures, or by pre-treatment of the minerals, although this method can require additional energy. The IPCC estimates that a power plant equipped with CCS using mineral storage will need 60-180 per cent more energy than a power plant without CCS.

The following table lists principle metal oxides of Earth's Crust. Theoretically up to 22 per cent of this mineral mass is able to form carbonates.

Earthen Oxide	*Per cent of Crust*	*Carbonate*	*Enthalpy change (kcal/mol)*
SiO_2	59.71		
Al_2O_3	15.41		
CaO	4.90	$CaCO_3$	-179
MgO	4.36	$MgCO_3$	-117
Na_2O	3.55	Na_2CO_3	
FeO	3.52	$FeCO_3$	
K_2O	2.80	K_2CO_3	
Fe_2O_3	2.63	$FeCO_3$	
		All Carbonates	

Leakage

A major concern with CCS is whether leakage of stored CO_2 will compromise CCS as a climate change mitigation option. For well-selected, designed and managed geological storage sites, IPCC estimates that CO_2 could be trapped for millions of years, and are likely to retain over 99 per cent of the injected CO_2 over 1000 years. For ocean storage, the retention of CO_2 would depend on the depth; IPCC estimates 30-85 per cent would be retained after 500 years for depths 1000-3000 m. Mineral storage is not regarded as having any risks of leakage. The IPCC recommends that limits be set to the amount of leakage than can take place. It should also be noted that at the conditions of the deeper oceans, (~400 bar, 280K) water-CO_2 mixing is *very* low (where carbonate formation/acidification is the rate limiting step), but the formation of water-CO_2 hydrates is favourable (a kind of solid water cage that surrounds the CO_2).

To further investigate the safeness of CO_2 sequestration, we can look into Norway's Sleipner gas field, as it is the oldest plant that sequesters CO_2 in an industrial scale. According to an environmental assessment of the gas field which was conducted after ten years of operation, the author affirmed that geographic sequestration of CO_2 was the most definite way to store CO_2 permanently.

"Available geological information shows absence of major tectonic events after the deposition of the Utsira formation [saline reservoir]. This implies that the geological environment is tectonically stable and a site suitable for carbon dioxide storage. The solubility trapping is the most permanent and secure form of geological storage."

Chemistry's Application in Industry

There are five principal methods whereby pressure vessels may be constructed, viz. by casting, soldering and brazing, riveting, welding and hollow forging. Oust vessels are usually of comparatively small dimensions, and for moderate pressures. Soldering and brazing methods are confined to small non-ferrous vessels. Riveted construction follows the same general principles which have been developed for steam-boiler practice, but due to recent marked advances in the art of fusion welding, riveting is fast being replaced by welded construction carried out under careful control.

Hollow forging is employed for vessels for the very highest pressures and severe operating conditions, as for example in the hydrogenation of coal, and the synthesis of ammonia and methyl alcohol.

In Great Britain the only official regulation for the actual design of steam boilers and pressure vessels is contained in the appropriate sections of the Board of Trade Instructions as to the survey of Passenger Steamships. These instructions are in respect of the general principles of design, and rules for computing the dimensions of the various stressed parts of pressure vessels, such as boilers, steam-pipes, air receivers and Diesel engine starting-bottles.

There are no corresponding rules for vessels for land service, but in so far as steam boilers are concerned, the Board of Trade rules are usually applied because all boiler explosions in this country are investigated by the Board of Trade under the Boiler Explosions Acts of 1882 and 1890.

These Acts contain clauses to the effect that *(a)* notice of all boiler explosions with full particulars must be sent to the Board of Trade within 24 hours, and *(b)* the word "Boiler" is defined as any closed vessel used for generating steam, for heating water or any other liquid, or into which steam is admitted for heating, steaming, boiling or other similar purposes.

The Factories Act of 1937 requires that every steam boiler for the generation of steam must be equipped with certain fittings, that boiler and accessories must be properly maintained, and that such boilers must be examined by a competent person every fourteen months. In practice, such examinations are carried out through one or other of the Boiler Insurance Companies.

The Factories Act of 1937 extends the principle of steam-boiler inspection (as required by the 1901 Factory and Workshops Act) to include practically every other type of steam-pressure vessel and air receiver, in that such vessels must be examined by a competent person once every twenty-six months.

There are certain other statutory requirements in regard to pressure vessels in works designated as Chemical Works under the Chemical Works Regulations, 1922 (S.R.O.No. 731, of 1922). Para. 5 requires that every still and every closed vessel in which gas is evolved or into which gas is passed and in which the pressure is liable to rise to a dangerous degree, shall have attached to it and maintained in proper condition a proper safety valve or other equally efficient means to relieve the pressure. Para. 6 of these Regulations deals with the provision of breathing apparatus and lifebelts, and this leads naturally to paras. 7, 8, 9, 19, 20, which govern the entry of workpeople into vessels, chambers, structures of any sort, and any room or place in which there is reason to apprehend the presence of a dangerous gas or the existence of an irrespirable atmosphere.

These regulations are of the utmost importance in regard to the design of plant and methods used in construction in regard to access, cleaning and repairs.

From the aspect of working pressure it is convenient to divide pressure vessels for the chemical industry into two groups.

1. Vessels for service at pressures not higher than about 1000 lbs. per sq. in. and wherein the "thin-cylinder" formula may be applied. This pressure limit may be raised if high-tensile alloy steels are used, since in such cases this working stress may be considerably higher than plain carbon steels and consequently the wall thickness is sufficiently thin for the " thin-cylinder " theory to be applied.
2. Vessels for service at pressures higher than 1000 lbs. per sq. in. and wherein the "thick-cylinder" or Lame's theory may be applied.

Each of these two groups may be again subdivided on the basis of the working temperature of the pressure vessel concerned, since if the working temperature of the stressed parts is within the region wherein " creep " occurs, the above principles of design based upon the thin-cylinder and thick-cylinder theories must be modified accordingly. The modification consists in substituting the safe working stress at the working temperature.

A pressure vessel consists essentially of a middle and two ends. The middle portion is practically always cylindrical in shape, and the ends may be either flat, dished, or hemispherical. The principles underlying the design of the shell or cylindrical portion will first be discussed.

Pressure Vessel: Copper Steam

The copper used in the construction of vessels should be in the hard condition, not annealed, and when forming a cylindrical shell subject to internal pressure, or when in tension not exposed to heating by furnace or flue gases, a maximum tensile strength of 5000 lbs. per sq. in. may be allowed. Superheated steam should not be used in vessels made of copper.

When copper is in the annealed condition, as a rule the maximum tensile stress should not exceed 3000 lbs. per sq. in.

The two most common forms of joint used in connection with copper plates are the brazed and the welded joint. Occasionally joints are clamped and brazed, riveted and brazed or riveted and soldered.

For the purpose of calculation the efficiency of an ordinary brazed or clamped and brazed joint should not be considered higher than equal to 50 per cent of the plate strength.

The design of riveted joints should be carried out on the same lines as for steel plates with steel rivets, but in the design of the copper joint, account requires to be taken of the effect on the design of the lower strength of copper.

Particularly, it is to be noted that the grip of the copper rivets holding the joint together will not be so great as in the case of steel rivets; consequently, on these grounds, a closer pitch will be desirable than in the case of steel rivets. Also, the copper rivet being weaker than a steel rivet of the same side, it will be necessary in a given pitch for the copper rivet to be of somewhat larger diameter than a steel rivet would be for the same pitch. The effect of this will be to modify the percentage strength of joint which would otherwise be obtainable.

Protection of Steam-heated Pressure Vessels against Overpressure

It is essential that steam-heated pressure vessels connected to steam boiler plant should be protected against dangerous overpressure if they are designed to work at a pressure lower than that of the boiler plant. This requirement is stressed in the recent Factories Act 1937, and may be illustrated by the following example:

> Assume that a reducing valve having an inlet bore of 1 in. diameter is supplied with steam at 120 lbs. pressure per sq. in. and is set to reduce the pressure in a low-pressure vessel to 10 lbs. per sq. in., and that no steam is being condensed in the vessel or discharged except through the safety valves. Also assume that the reducing valve has become deranged and stuck fast in the full open position, thereby allowing high-pressure steam to enter the low-pressure main. The problem to be solved is what should be the safety-valve area under such conditions to ensure that the pressure in the low-pressure vessel is not increased above 10 lbs. per sq. in.

The problem can best be studied from the laws governing the flow of steam through orifices, the supply pipe being considered as an inlet orifice to the low-pressure vessel and the safety valves as outlet orifices therefrom.

The Hollow Forging of Pressure Vessels for Very High Pressures.

The increasing demands of recent years for large high-pressure boiler-drums and chemical reaction vessels has led to the development of hollow forging of such vessels because the wall thicknesses are too great to permit of riveted construction, and in the case of chemical vessels, the materials are frequently of such a composition that welded construction is inadvisable.

Such hollow-forged vessels may frequently have a finished weight of 60-70 tons, which necessitates an ingot from which to forge the vessel of 160-180 tons, and this in turn has led to marked developments in ingot moulding.

In the manufacture of hollow forgings until recently the usual practice was to anneal the ingot after stripping it whilst still hot from the mould, afterwards allowing it to cool very slowly. When cooled, a hole was trepanned through the centre, the object of which was to produce a core of metal which could be closely examined, and also to remove the less pure material from the axis of the ingot.

The next operation was to enlarge the hole by an hydraulic pressing operation after the ingot has been reheated to the necessary forging temperature, and this produces a long tubular forging. Subsequently the drums are normalised, the outside is rough machined and the interior bored. After reheating the ends are closed for manhole openings, in the case of boiler-drums, or are reduced and then flanged in the case of chemical reaction vessels.

It will be realised that the above process is a lengthy and costly one, and attempts have recently been made to reduce both the labour costs and the production time. The first of these consists in hot piercing or punching whereby a hole is made through the ingot without cooling it. In this process, the feeder head at the top and the surplus metal at the bottom of an ingot is removed by a large cutter, driven through the ingot by means of an hydraulic press. The remaining body of the ingot is then placed vertically beneath a press, and a tubular punch driven axially through, thus removing a core from the centre of the ingot.

Another method recently adopted involves the casting of a hollow ingot, and it has several great advantages. Firstly, economy in raw material is secured, and secondly considerable reduction in time of manufacture is obtained.

Until recently, large pressure vessels were available only in the carbon steels, but now large ingots can be obtained either by open hearth or electric furnace process in alloys such as 3 per cent, and 6 per cent, chrome molybdenum, and the 25 per cent, nickel-chrome molybdenum types. The subsequent heat treatment of such large masses of alloy steel is a difficult and costly process, but in view of the special features of such alloys, manufacturing costs are not prohibitive.

Uses of Steam-pipes

It is used in nearly every branch of chemical engineering, and it is of importance., therefore, that strum supply pipework be well designed, adequate in size or the purpose desired, well covered to prevent unnecessary heat losses, properly drained to prevent possibilities of damage by water hammer and to provide dry steam at the point of use.

With the introduction of mild steel pipework at the beginning of the present century, the manufacture of cast-iron and wrought-iron steam-pipes steadily declined, and within the limits of susceptibility of this material practically all steam-pipe installations are now of mild steel throughout. For extreme temperatures at high pressure alloy steels are imperative, but since these conditions are of recent growth the technique is by no means stabilised.

J. A. Alton gave an interesting paper on modern high pressure-high temperature pipework, a summary of which is as follows:

> Pipework for severe operating conditions must he considered from three main aspects: (a) the raw material of the pipe, (b) the method of manufacture, and (c) the working conditions, both as regards

pressure and temperature of the gas or fluid being conveyed, and the arrangement of the pipe.

At the higher temperature ranges, the "creep" of steel must be taken into consideration; and if the maximum life of the pipework be taken as twenty years, then it is reasonable to assume that the actual time which the plant will be under service condition is, say, 100,000 hours.

Since the permissible creep in this time may be taken as 1 per cent., it follows that the allowable creep per inch per hour is one-tenth of one-millionth, i.e. 10^{-7} ins. Taking 60 per cent of this creep as the maximum permitted to allow for contingencies, the following stresses are suitable for 0.17 per cent, carbon steel.

In using the foregoing formula and data in determining pipe dimensions it must be borne in mind that manufacturing difficulties do not permit of 0.15 per cent, carbon steel tubes with wall thicknesses necessary at pressures approaching, say, 1400 lbs. per sq. in. at the maximum working temperature given, viz. 900°F.

Recourse must then be had to alloy steels, and since such steel for pipe fabrication has to be capable of being worked at high temperatures in the manufacturing process without subsequent elaborate heat treatment, and of being welded for flanges, etc. without detriment to its mechanical properties in the process, great care is necessary in selection.

Steam-pipe Connections and Flanges

To connect saturated steam pipework at pressures not exceeding 150 lbs. per sq. in. and pipe diameter of 2 ins., it is satisfactory to use standard screwed bronze unions, screwed tees, bends or sockets, but above those pressures and diameters it is advisable to use flanges, and flanged tees or bends.

The dimensions of flanges and flanged fittings, and the method of securing flanges to the tube, depends largely upon the working pressure and temperature. To avoid carrying an excessive variety of flanges it is advisable to standardise on as few different sizes as the plant will permit.

Flanges may be fixed to pipes by welding, screwing, expanding or riveting, or by a combination of two or more of these methods. Expanding alone is definitely not recommended for even the lowest pressures.

Welded-on flanges can be supplied for all sizes, temperatures and pressures. The riveted flange is not recommended for high temperature and it is difficult to fit with sizes less than 7 ins. diameter.

Screwed flanges are suitable for pressures up to 200 lbs. per sq. in., and screwed and expanded flanges for pressures up to 350 lbs. per sq. in. at temperatures not exceeding 650°F.

Regarding cost, the riveted flange is the most expensive, and the screwed-on flange the cheapest type.

In the case of screwed- and expanded-on flanges it is usual for the expanding to be done with proper appliances at the tube-maker's works.

It is important in such cases to note that the pipe is screwed with a vanishing thread, so that the pipe is not unduly weakened by the expanding process at the end of the thread. If such pipes are likely to be subjected to torsional stress it is well to reinforce the attachment by a light weld fillet at the back of the flange.

Such flanges should be welded to the pipe which is a type which can be well recommended. The chief features of this welded connection are:

1. The main weld is a butt-weld and is consequently either in tension or compression but never shear, and consequently a minimum of weld material is required.
2. The bending stresses on the butt-weld arc reduced to a minimum.
3. In subsequent facing of the flange none of the weld-metal is removed.

The butt-weld should be by the metallic; arc process, and the fillet either by metallic arc or carbon arc process.

When the steam pressure exceeds 350 lbs. per sq. in. and the temperature 750°F, great care must be exercised in choosing the type of flanged connection from pipe to pipe, and pipe to fittings or valves.

The design of pipe flanges for very high-pressure steam lines has naturally received the attention of the British Standards Institution, who in 1932 issued a tentative specification for flanges of pipework and fittings to be used for working pressures from 900 up to 1400 lbs. per sq. in. at temperatures up to 800°F, but extended by a reduction in the working pressure to 900 lbs. per sq. in. to a maximum of 900°F. In issuing this specification the Committee pointed out that the recommendations were in advance of practice and knowledge and the tabulated data were thereupon issued as a *tentative* standard only.

For the very highest pressures there are three forms of welded high-pressure pipe connection which have proved successful in practice. The first was developed in America and is known as the "Sarlun" joint. The Sarlun joint is a development of the "Sargol" joint, which was in effect the old well-known "Van Stone" joint. In it the collars are the same as in the Sargol joint, but the collar is turned down circumferentially, thus leaving on the face side of the collar a raised V-shaped circumferential ring. This being thin, requires much less heat in the welding operation. The faces of the collar are scraped and made steam-tight before welding.

The second form of seal-weld joint is the " Corwel", developed in Great Britain. This joint uses a loose flange of similar shape to the " Sarlun," but the collar is produced by raising a corrugation on the end of the pipe and rolling this back, thus producing a double or folded collar. The face is machined smooth, and a heavy weld is used without difficulty.

The bolts used for the Corwel joint are nickel-chrome molybdenum steel and are not formed with head but are screwed fine thread right through.

High-temperature Steam Experience at Detroit

The authors report experiments on two high-pressure high-temperature steam installation, one at the Trenton Channel power-station and one at the Delray powerhouse No. 3. The first experimental plant at Trenton started up in 1929 comprised a separate oil-fired superheater and piping system designed to operate with steam at 1000°F, and later at 1100°F, and this was operated for two years to give data for the second stage of the experimental work which comprised the installation of a 10,000 kW. steam turbine with necessary equipment for working at 1000°F. The working pressure throughout was limited to pressures not higher than 400 lbs. per sq. in., and it was considered that the higher temperature, without the added complication of very high pressure, would provide the information most desired.

Description of Equipment: The Trenton equipment comprised an oil-fired superheater supplied with 400 lbs. 700°F. steam from the power-station main header. The superheater was of the radiant type, and was capable of adding 400°F of superheat to 6000 lbs. of steam per hr. from the initial temperature of 700°F. The high-temperature piping system consisted of a valve, eight flanged joints, and a de-superheater in order that the steam used in the experiments could subsequently be used in the house turbine at 700°F.

The superheater was made of low-carbon steel tubing in the section where the steam temperature did not exceed 800°F, and the remainder of the superheater and the high-pressure piping system were made from stainless steel with analysis C 0.06-0.09, Cr 17-20, Ni 7.10, Si 0.5. The steam piping is 5′ diameter, outside with thick walls. The loose companion flanges, valve and desuperheater castings were made of similar material to the above, except that the carbon was increased to 0.25 and the silicon to 2.0-2.5 per cent.

The flange bolts were made from chrome-tungsten vanadium steel. The pipe ends were provided with Van Stone Sarlun laps, the faces of whitch were provided with a serrated finish to provide data for gasketed as well as seal-weld joints. All piping joints were constructed to the 600 lbs. A.S.A. standard. The original line has been somewhat modified by the addition of experimental joints of heavier construction.

Numerous gasket materials were tried, and the best was found to be 3-in. Monel metal in plain sheet form.

Several different classes of low-alloy stools were installed for experimental purposes in both Trenton and "Delray installations. Apart from savings in material cost, such low alloys are more readily fabricated. Among the castings tried were 46 per cent. Or 1 per cent. W 0.6 per cent. Mo; a 5′ outside diameter length of 4.6 per cent. Cr 1 per cent. W; a valve body of Hadlields Era 131; medium- and high-carbon calorised stool together with a nickel-chromium molybdenum steel adopted by the turbine manufacturers for all parts of the turbine subjected to high-pressure steam.

A few tubes of medium- and high-carbon calorised steel were installed in the Trenton superheater, arid 0-35 C steel tubes coated with aluminium were installed in the middle section of the Delray superheater, together with a like number of coated tubes of 18.8 and 12.14 per cent, chrome-nickel austonitic steels. Metal gaskets were used for all unwelded joints, and the flanges were made of low-chrome nickel stool.

Results of both physical `and microscopic examinations made on various materials after service from the two installations have shown slight change after service under the experimental conditions, but in no case have the changes been sufficiently serious as to warrant replacement of affected parts. Six different classes of alloys were examined.

Three different sections of 18-8 material removed from the Trenton installation showed evidence of carbide segregation at the grain boundaries; and each successive sample revealed widening of these boundaries. The physical properties appeared to indicate slow embrittlement, but such changes were not considered to cause concern as to their continued service. The authors recommend the development of special alloys to avoid the chromium-carbide segregation.

The calorised superheater tubes and headers of medium- and high-carbon steel were removed from the Trenton superheater after 3821 hours' service, of which 3305 hrs. were above 1000°F, and stresses due to all circumstances as high as 3000 lbs. per sq. in. Approximately 10 per cent, of the calorised surface coating had disappeared and the remaining 90 per cent, was quite brittle. The inside of the tubes was coated with 0.04-0.08 thickness of magnetic oxide, and measurements showed creep rates of 2 per cent, per 1000 hours' service.

The Dolray calorised and sprayed superheater tubes were not removed for inspection, but the external surface was apparently in quite good condition.

A valve at the outlet of the Trenton superheater made from 0.4 per cent. Or 1 per cent. W was removed for examination after 12,995 hrs., 10,818 of which were at a temperature of 1100°F The valve was 600 Ihs. A.S.A, construction, and the test results show that the material of the valve body did not undergo embrittlement, but that the long service at high temperature tended to increase the toughness and ductility. Experience with Nitralloy showed that it is unsuitable for valve parts.

Samples of the nickel-chrome molybdenum steel from the throttle valve of the turbine after 3800 hrs., 1450 of which were at 1000°F, showed but little change in physical properties.

Various bolting materials were tried with varying results. Bolts made from this material were heat-treated by oil quench from 1650°F followed by 1100°F tempering, and the threads were cut after heat treatment to prevent the formation of cooling cracking the thread roots.

Method of Pipe Jointing

Many and varied troubles were experienced with pipe jointing, but from experience gained it was found possible to construct two joints which successfully withstood 16,000 hrs. at steam temperatures of 1000°F. The authors state that full weld-joints appear to be the answer to the high-temperature problem.

The successful 5-in. bolted flange joint was of the Van Stone type with serrated flanges of KA2 material. The bolt stress was limited to 10,000 lbs. per sq. in., and one joint of this type had flanges as 900 lbs.

No material dimensional changes were noted in the 8-in. Delray pipe stressed to 6400 lbs. per sq. in., or in any of the fittings, and measurement of a section of 51 in. outside diameter chrome-tungsten tube after service in Trenton likewise showed no change after 7468 service hours.

In a concluding paragraph, the authors state that the use of steam equipment up to 1000°F is entirely feasible, but this is qualified by the statement that if such operation demands existing high-priced alloys, it is scarcely workable. Further metallurgical research will no doubt, however, produce durable materials at a much lower cost, when such high temperatures will be justifiable.

Bends, Expansion Bends and Joints

Bends: In designing large bends for steam mains three rules should be observed: (a) all bends should be formed on long lengths of tube wherever possible, (b) all bends should have at each end a straight portion not less than one and a half pipe diameters, and (c) all bends should be of as large radius as possible. In some installations, however, the layout may necessitate the use of bends of short radii, and in such cases creased or corrugated bends may be used.

It must be realised that creased and corrugated bends cause largely increased pressure drop and their use is not recommended unless the layout renders this imperative. Where bends of still shorter radii are necessary, these can be formed by guesting, i.e. wedge-shaped pieces are cut from that side of the tube which is to form the inner curve of the bend and the edges of the opening are bevelled. The tube is then bent, forcing the cut edges together, and these are then welded. As an alternative to such gusseted bends cast steel elbows may be employed.

With small radii and very thick bends, the bending process may thin the outside wall of the bend by 30 per cent. Creasing the inner wall makes it possible to bend pipes of the dimensions given to a smaller radius without thinning the outer wall, and such creasing definitely increases the flexibility of the pipe. Creasing a bend enables it to be made in large diameters.

The fully corrugated pipe can be made to smaller radii even than the creased pipe. They are more flexible. No case of corrosion in the walls of the corrugation has come to the notice of Mr. Aiton, who has had very wide experience of this class of work.

Expansion Sends: Provision for expansion must be made in the design of every steam main, and this is best done by arranging a sufficient number of ordinary bends to allow for all movement. Where it is not possible to do this, expansion bends of this type may be used. In arranging the pipe movements, care should be taken that branch connections to the main do not cause restriction of such movement, otherwise serious damage to the main may result.

Sliding expansion joints of the stuffing-box type are used where space does not permit of the use of expansion bends, but such joints should only be used for low pressures and temperatures.

Great care is necessary in erecting and packing such joints, as leakage from the gland will otherwise result. Copper expansion bellows are sometimes used, but they are not recommended for pressures higher than 10-20 lbs. per sq. in., or where torsion or excessive movement may possibly arise.

Erection of Expansion Bends

To use an expansion bend to the best advantage it is necessary to spring it during erection. In doing this, care must be taken to prevent any movement of the main at the anchors, since this would mean the loss of so much of the cold springing and consequently an increase in the stress imposed on the bend in the working condition if it is still to take up the same amount of expansion movement.

Pipe anchors in suitable positions are essential for the satisfactory working of all types of expansion bends or expansion joints, and even where these are not employed anchors are frequently necessary to regulate the movement of the main. In long mains, guides to maintain the alignment are necessary both for expansion joints and expansion bends—for expansion joints because any misalignment will probably cause the joint to jam; for expansion bends because the end load necessary to produce the safe travel of the bend may be greater than the adjacent pipes can supply without themselves bending.

Expansion of Steam Pipes

Formula for calculating Expansion: The amount of expansion which will occur on any steam main can be calculated from the formula:

$$E = KLf$$

where E = amount of expansion, in ins.

L = length of piping under consideration, in ft.

t = maximum temperature variation in °F,

and K lias one of the following values, depending on the range of temperature.

The total variation in temperature should be determined basing on a minimum temperature appropriate to the local conditions.

Branches and Connections to Mains

For low pressures, i.e. up to 150 lbs. per sq. in., and for pipes below 5 ins. diameter, screwed tees are quite suitable for branch connections. For higher pressures and larger pipes it is usual to take branch connections from flanged tees or from riveted or welded-in branches on the main.

In attaching welded branches most makers cut out the opening in the main or manifold by oxyacetylene cutter and then shape the contour of the branch piece to the pipe opening. The connection should then be welded with a good fillet. When possible branches should be welded on at right angles to the main, as the large flat surfaces formed at the junction of an angle branch with the main are subject to a much higher stress than the cylindrical parts. Bosses welded on are often used in place of branches for small connections to mains. They are either tapped to receive the screwed end of a pipe or tapped for studs for flange connection.

If desired and over 10 ins. diameter riveted branches can be used where the pressure and temperature are low. Such branches should be placed near the ends of the main to permit of efficient holding of the rivet-heads while riveting. The riveted branches should be of pressed steel of similar design to boiler nozzles or standpipes, and should be of thicker gauge than the standard pipe size so as to allow for thinning in flanges, and give sufficient landing edge for caulking.

Steam Separators and Receivers

Complication in separator design is not desirable; it usually adds little to the efficiency and often causes unnecessary loss of pressure.

Steam separators depend for their powers of extracting water from steam on the following principles:

1. Sudden alteration of direction of flow.
2. Reduction in velocity of flow.

Separators are designed to utilise both of these principles. Separators should have hand holes in the body where the drain brandies are not sufficiently large to take their place.

Steam Receivers: Riveted steam receivers are suitable for steam pressures up to 500 lbs. per sq. in. working pressure at temperatures not over 650°F The receiver shell is formed from one plate, the longitudinal joint being butted, fitted with double butt-straps, and electro-welded at each end to ensure steam tightness. Each receiver end is made from a single plate pressed to shape in one heat and machined to ensure a good fit in the shell, which is bored out to receive the ends. The ends are of special form, giving the maximum strength combined with a flat large enough to accommodate the usual elliptical manhole for access to the interior. Standpipes for this type of receiver are solid forged, and double riveted to the receiver.

For working pressures above 500 lbs. per sq. in. at temperatures exceeding 650°F solid forged receivers are recommended. On these the standpipes are solid forged and are welded to the receiver.

Supports for Steam-piping

The supporting of steam-pipes is closely connected with the provision for movement of expansion and contraction, but the supports proper should, as far as possible, permit free movement of the pipes in any direction, and any necessary constraint should be provided by anchors.

Where pipes are subject to vertical movement or to vibration, spring supports should be employed to avoid excessive stresses which might be set up in the pipes by the use of too rigid supports. In designing such supports the selection of springs of the proper strength is of importance. If the springs are too weak they will allow undue vibration to take place, whilst if too rigid they will not relieve the stresses in the pipeline. The spacing of supports is largely determined by the position of floor girders, joints, columns or other structural features, but supports must not be placed too far apart. As far as possible, pipe joints should be placed close to the points of support.

Although in many cases pipelines can be allowed free movement, subject only to constraints imposed by the fixing of their end flanges to the units which they connect, it is sometimes necessary to introduce fixed anchors to minimise the stresses which might otherwise be set up at certain points of the system. The location of anchors must be very carefully considered in relation to the system as a whole.

As auxiliaries to fixed anchors, limiting or guiding anchors are sometimes required to prevent movement in some particular direction or to maintain the correct alignment of the pipes under the loads imposed by expansion.

Drainage of Steam Mains

The proper drainage cf steam pipework and receivers is of the utmost importance, and such drainage should include provision for the immediate removal of the water formed by condensation, and for the water which accumulates in the pipes after steam is shut off the main.

It is to be noted that water will not readily flow against the direction of steam flow even if the pipes are given a steep inclination. A normal steam speed of, say, 100 ft. per sec. is comparable to a wind velocity of 60 m.p.h., which is likely to cause water to back up against the direction of flow of the pipes.

Where possible, the boiler junction valves should be the highest point of the range, otherwise the drainage of a branch steam-pipe to a boiler which is out of action is difficult and the presence of water in this pipe is particularly liable to set up water hammer. If necessary and possible, the boiler junction valve may be raised by inserting a standpipe between the boiler and the valve.

To unsure that pipes can be cleared of water before steam is admitted, it is necessary that there be a drain and an efficient steam trap at the lowest part of each section that there be also means for admitting air to each section, preferably at the highest point.

All drain connections should be of ample bore to prevent choking, and should be made in the mains at such points as at valves of smaller diameter than the main, expansion bends in vertical planes, the foot of each rising bend and all similar positions. It is specially important that branch pipes rising from the main and provided with stop valves next the main should each have a drain connection to prevent water from accumulating when the branch pipe is not in use. The occurrence of such danger-points can often be avoided by slight changes in the design.

To take drain connections from mains either bosses welded to mains, drilled and tapped to take the screwed end of the drain pipe, should be used, or branches welded on and flanged to match the flange on the drainpipe. If more water is expected than could be satisfactorily discharged by a drainpipe connected to the main, a drain pocket of proper size welded to the main, flanged and fitted with a blank flange or with the end closed, should be employed, the blank flange or closed end being fitted to take the drain connection.

Design of Steam Traps

There are many designs of steam traps on the market, and care should be taken to select the proper type for a particular purpose. Traps may be classified into two main types, viz. lifting and non-lifting. The latter is used when the condensate is discharged at a level lower than the trap and where the discharge of the condensate imposes no back pressure upon the trap. Lifting traps are used to raise the condensate to a level higher than the trap, in which case the trap must frequently be designed for a considerable back pressure.

Every trap should be provided with some means of testing its condition while at work, and with a bypass to permit of its being removed for repairs.

Many manufacturers specialise in the manufacture of all types of steam traps, and when purchasing such appliances the fullest information should be given in the specification and enquiry in order that the most suitable type is supplied.

Steaming at Reduced Pressure

Steam at reduced pressure is frequently required in many departments of a chemical works. It is usual to supply steam at high pressure to the department concerned, and reduce, close to the point of use, thereby keeping the steam supply pipe as small as possible.

The principle of most reducing valves is the balancing of a double beat valve by assisting the low-pressure side of the valve by a spring, which can be varied in strength for various reduced pressures. In this way the pressure in the reduced main is automatically controlled, since if the pressure in the reduced side rises beyond that arranged for, the double beat valve closes and steam is not again admitted until the reduced pressure falls.

Relay reducing valves are desirable especially with widely fluctuating loads to prevent hunting of the valve arid wide variations in steam pressure.

Formation of Vessel Shells

If a thin circular cylinder is subjected to internal fluid pressure, a tensile stress arises in the walls of the material, and this is usually described as the "hoop stress". When the thickness of the cylinder walls is small compared with the internal diameter of the cylinder, it is assumed that the hoop stress is uniformly distributed throughout the cross-section of the material.

Pressure Vessels: Construction

The general practice of riveting in chemical pressure-vessel work follows closely the technique long developed for steam-boiler construction. It is of importance to note that apart from failure by leakage a riveted joint can fail only in certain well-defined ways:

1. Failure of rivets in shear.
2. Tearing of plate through a line of rivet-holes.
3. Combined tearing of plate and shearing of rivets.
4. Lap rupt 111 × 3.

These four possibilities form the basis for the design of riveted joints:

1. Is accounted for by providing an adequate section of rivet-material in shear;
2. By providing an adcquate section of plate to resist the tearing process;
3. By taking account of the weakest section of the plate and adding to its strength, the resistance of the rivets which would have to shear in order to allow the tearing to take place; and
4. By providing adequate material between rows of rivet holes and between the holes and the plate edges.

In general, the efficiency of a riveted joint may be defined as the ratio between the minimum calculated resistance to failure, and the ultimate tensile strength of the solid plate, and this ratio is normally used in the calculation of the working pressure. It should be noted that it is the ultimate tensile strength of tin; plate material which provides the criterion of efficiency are riveted joints, and, so far as pressure vessels are concerned, this criterion has been found by experience to be entirely satisfactory.

Riveted construction, however, does not permit of good design in the case of many chemical pressure; vessels: first, because of the interference of riveted joints with a smooth interior; second, the stresses set up in the shell and rivets in the riveting process often lead to rapid local corrosion; third, because many metals do not lend themselves to riveted work.

The following rules on riveting of boilers are taken from the Board of Trade Rules by permission of the Controller of I.T.M. Stationery Office, and arc given as a guide; to good practice in the riveting of pressure vessels.

Note that these rules apply to "wlired" pressure vessels, and for vessels of the unfired type the working pressure may be increased by 10 per cent above that found by these formulae for fired vessels.

Distances between Rows of Rivets and between Rivets and Plate Edges:

1. The clear space between a rivet hole and the edge of a plate should not be less than the diameter of the rivet hole, i.e. the centre of the rivet hole should be at least U diameters distant from the edge of the plate.
2. In zigzag riveted joints, whether lapped or fitted with butt straps, in which there is an equal number of rivets in each row.
3. In chain-riveted joints, whether lapped or fitted with butt straps, in which there is an equal number of rivets in each row, the distance between the rows should not be less than 2*d*.
4. In zigzag riveted joints in which the number of rivets in a row is one-half of the number in an adjacent row.
5. In chain-riveted joints in which the number of rivets in a row is one-half of the number in an adjacent row.

The distance between rows in which there are the full number of rivets should be not less than $2d$. Where d is diameter of the rivet holes.

Methods of Calculating the Strength of Riveted Joints: The percentage of strength of a riveted joint is found from the following formulae (I), (II), (III): (I) and (II) are applicable to any type of joint; (III) is applicable only to joints in which the number of rivets in inner rows is double that in the outer row. The lowest value given by the application of these formulae is to be taken as the percentage of strength of the joint.

(I) Percentage of strength of plate at joint as compared with solid plate – 100 $(p - d)\ P$

(II) Percentage of strength of rivets as compared with the solid plate – 100 $(S_2 \times a \times n \times C) \sim S_1\ x{\sim}p \approx x{\sim}T$

(III) Percentage of combined strength of the plate at the inner row of rivet holes and of the rivets in the outer row = 100(p–2d), 100$(S_2 \times a \times C) \sim p\ S_1 \times p \times T$

where p = pitch of rivets at outer rows in in.,

d = diameter of rivet holes in in;,

a ~ sectional area of one rivet in sq. in.,

n = number of rivets which arc fitted in the pitch p,

T = Thickness of plate in in.,

0 = 1.0 for rivets in single shear as in lap joints,

C ~ 1.875 for rivets in double shear as in double butt-strapped joints,

S_1 = minimum tensile strength of plates in tons/ sq. in.,

S_2 = shearing strength of rivets, which is taken generally to be 23 tons/sq. in., and may be 85 per cent, of the minimum tensile strength of the rivet bars.

Circumferential Seams

1. The riveting of the seams joining the end plates to the cylindrical shell shall be not less than 42 per cent, of that of the solid plate. Where the shell plates exceed 1 in. in thickness the seams connecting the shell plates to the end plates are to be double riveted.
2. The circumferential seam at or near the middle of the length of single-ended boilers should have a strength of joint not less than 60 per cent, of the solid plate. The inner circumferential seams of double-ended boilers should have a strength of joint not less than 62 per cent, of the solid plate. In any case, there shall be three rows of rivets.
3. The circumferential Seams in the shell of a vertical boiler shall be not less in strength than 42 per cent, that of the solid plate. When the seams are not complete circles, they shall be double riveted.

Welded Construction of Pressure Vessels

Before discussing the various rules for welded construction of pressure vessels it is considered desirable to describe in detail the various methods of welding followed by notes on design.

Pressure Vessels: Fabrication

The welding processes used in the manufacture of pressure vessels, fabricated plate work and accessories may be broadly classified as follows:

1. Forge and hammer welding.
2. Electric resistance butt welding.
3. Atomic hydrogen process.
4. Metallic arc fusion welding.
5. Acetylene welding.

Forge and Hammer Welding

The forge and hammer welding method of pressure-vessel construction is closely allied to the oldest known method of welding, and has been used for generations in the manufacture of furnace and flue tubes for Lancashire and Cornish boilers. This method has recently been developed, chiefly in Germany, for the fabrication of large pressure vessels, and the following description of the technique adopted by Messrs. Thyssen of Mulheim Ruhr is of interest in this connection.

The pressure vessels are made from standard qualities of steel plate, and for the cylindrical portion, the plate is first put through rolls to obtain circular shape. The welding of the longitudinal seam is carried out with a water gas flame of a reducing nature, and the edges of the plate being welded are worked by means of rolling at welding temperature. After completion of the longitudinal seam, this is then closely examined, and if satisfactory the vessel is annealed and an internal hydraulic; pressure of 50 per cent, above the desired working pressure is applied. The ends of the vessel are then closed in, and manhole and other openings fitted.

Upon completion, the vessel is then highly stressed by application of an hydraulic pressure of three or four times the working pressure, and finally the vessel is heat-treated to relieve internal stresses.

During the past twelve years, Messrs. Thyssen have supplied more than 1250 such vessels, chiefly in the form of steam drums for water-tube boilers, and in the majority of eases the working pressures were in excess of 500 lbs. per sq. in.

Electric Resistance Butt Welding

For joining of parts of tubular structures this method offers many advantages, and it can be used with great facility in the construction of elements of tubular heaters and heat exchangers.

It has the advantage that when the weld is complete, there has been no addition of metal from any outside source. The method of operation is to place the tubes to be welded one into each of two clamps which form the electrodes of the machine. One of the clamps is stationary and forms part of the rigid structure of the machine, and the other forms a sliding head attached to a mechanically operated ram capable of either slow or rapid movement.

Current flows between the two clamps through the tubes, and when the ends of the tubes meet a resistance is set up due to discontinuity of the structure. This resistance raises the temperature of the metal at the ends of the tube, a certain amount of metal is flashed off, and when at the right plasticity an upset or forge blow is imparted to form the weld.

The current used is supplied through a transformer, the secondary windings being tapped to give voltages ranging between 5-8 volts. The cost of equipment and of power consumption in this process limits the size of work to tubular structures in which the cross-sectional area to be welded does not exceed about 30 sq. ins.

Atomic Hydrogen Process

The use of atomic hydrogen is one of the later developments in the field of welding and appears to have certain advantages. Although an electric arc is used, its application is indirect, and up to the present covered electrodes have not been utilised to any extent. It may be looked upon as supplementary to metallic arc welding rather than as a competitive process.

An arc is maintained by an alternating current between two tungsten wire electrodes and a stream of hydrogen gas is introduced around the arc. The temperature of the electric arc is sufficiently high to cause dissociation of the hydrogen molecules, and the atomic hydrogen on leaving the zone of the arc recombines to form again hydrogen in the molecular state.

The latter process is a reversal of the heat input from the electric arc, energy is given up, and provides the heat required for welding.

Such a process may appear to be a complicated method of providing the heat required for welding, but by this means it is possible to produce a flame of high temperature composed of a single gas without the usual combination with oxygen, whilst, in addition, the welding can be carried out in an envelope of burning hydrogen so that the metal constituting the weld can be maintained free from atmospheric contamination without a slag coating.

Metallic Arc Fusion Welding

In view of the importance of this method in the fabrication of pressure vessels, the subject will be dealt with at length.

When the early development of welded pressure vessels began, electric arc welding offered great possibilities. An automatic machine had been devised for welding cylindrical containers made from thin steel plates. This machine worked well with bare wire electrodes, but the class of weld which could be made was unsuitable for pressure vessels since metal deposited in the weld from bare electrodes contains a number of major defects, such as incomplete fusion with parent metal, slag inclusions and porosity.

Apart from the presence of such defects, the use of the bare electrode results in considerable atmospheric contamination of the weld metal, this contamination being exaggerated by the instability of the arc when using electrodes of this type. The atmospheric contamination causes oxide and nitride inclusions in the weld metal which tend to make the metal brittle, even if present only in minute quantities.

The weld metal must have physical and chemical properties similar to those of the parent metal, and means must be provided to replace in the weld metal the carbon, silicon and manganese lost due to volatilisation of these elements in the arc. Coated electrodes, therefore, have to be used, and machines have to be developed capable of feeding such electrodes without stopping and starting at short intervals for electrode replacement.

The manufacturer has to consider vessels up to 80 ft. in length, in which long longitudinal seams are a necessity. Means have been found for automatically feeding long-coated electrodes to provide a continuous weld.

The advantage of such a method of working, combined with constant, mechanically regulated short arc conditions, will be easily appreciated. Weld metal can be laid down within definite temperature limits so that the class of structure can be controlled throughout the welding operation.

The electrode consists of two portions, the steel wire and the coating attached thereto.

The wire is delivered in coils and the first operation involves the cleaning of its surface and straightening and cutting it into the 8-ft. lengths suitable for the automatic welding machines. In the second operation projections are stamped by the machine on to the wire at a regular pitch.

These projections just protrude through the electrode mating, so that electrical contact can be made with the contactors of the feeding device on the automatic machines. In order that the automatic, welding process is not interrupted, the electrodes are made so that the one can lit into the other, thus providing what is essentially a continuous electrode. After the stamping operation, the wire is fed to the machine which extrudes the coating, and applies a carefully controlled thickness of the coating to the wire.

The Welding Technique: The deposition of weld metal by means of the metallic arc involves a close study of three main variables:

1. The electrical input to the arc;
2. The speed at which the electrode travels along the seam being welded; and
3. The shape of the groove formed by the abutting plates.

The quality of weld metal deposited from an electrode depends largely on the amperage and voltage of the arc. The second variable may be considered as follows: with given electrical conditions and an excessive speed of travel of the electrode along the seam, the heat generated in the vicinity of the arc will not be sufficient to fuse the sides of the groove being welded, and at the junction between the weld metal and the plate there will remain a thin film of non-metallic material, for instance slag, which will constitute a definite weakness at the joint. Time is naturally an important factor in raising the temperature of the parent metal to the degree necessary.

If the speed of the electrode is too low, three conditions can follow. First, the run of weld metal deposited is of too great a thickness so that subsequent recrystallisation by the succeeding run is confined to a relatively small depth of the metal. Secondly, the heat generated in the vicinity of the arc causes excessive fusion of the plate, and pockets enclosing slag are formed. Thirdly, the slag arising from the electrode coating, instead of being controlled to freeze behind the arc, will flow in front of the electrode and become overlapped by the weld metal.

Having once, therefore, determined the electrical input to the arc for good-quality weld metal, the speed of travel of the electrode must be considered in relation to the shape of the groove in the plate.

A groove, having the U-shaped section is almost universal for heavy plate because the change in the width of the groove after each layer of metal has been deposited is almost negligible, and because the almost vertical line of fusion is admirably suited to X-ray inspection.

The minimum dimensions of the groove arc controlled by the electrode used. The narrower the groove, the less the amount of weld metal necessary and the cheaper the fabrication. The factor of primary importance which determines the minimum width of groove is the first layer of weld metal deposited. If the groove is too pointed at the root of the weld, inadequate penetration at the base of the groove will follow.

Regarding size of electrodes, it might appear that the most economical procedure would be to use electrodes of large cross-section, thereby filling a deep groove in two or three runs. It sometimes happens in such cases that cheapness and speed do not go hand in hand with the soundness of weld in the manufacture of pressure vessels.

The four major considerations are:

1. Distortion of the vessel due to unbalanced contraction where large amounts of weld-metal are laid down;
2. The retention of a coarse columnar structure, because of the restricted depth to which the refining effect of the superposed run of weld-metal can penetrate;
3. The trapping of slag in the weld-metal due to freezing before the slag has floated to the surface of the weld run;
4. High current density necessary with large electrodes produces excessive porosity.

The multilayer weld is most attractive. The comparatively thin layers of weld-metal permit the full flotation of slag before freezing, and with suitable electrode coatings a protective slag with a low specific gravity can be ensured. Whilst the current density can be maintained sufficiently high to ensure good fluidity of weld-metal and complete fusion, the smaller section electrodes do not necessitate current densities of the class which bring about porosity.

Attention must also be paid to the disturbed area in the parent metal immediately on each side of the weld. There are many practical difficulties which prevent the normalising of the finished drum. It is not possible to heat the finished drum to a temperature of the order of 900°C to recrystallise the material.

In addition, considering the weld-metal, this recrystallisation would be undesirable because the rate at which the whole mass of the drum could be cooled through the upper 400°C. of the temperature curve would be such that a grain structure of very much coarser type would result in the heat-treated weld itself, and the laminar structure of the plate would lose the fibrous nature which makes it particularly suited to carry circumferential stresses, such as the hoop stress, etc. to which the drum is subjected in service. The only theoretical reason upon which recrystallisation in such a vessel can be advocated is to remove the disturbed area in the parent metal immediately adjacent to the point at which fusion between the weld-metal and parent metal takes place.

In discussing the structure of the parent metal adjacent to the weld, reference has been made to the important question of heat treatment. In considering this question, it is desirable to bear in mind the class of vessel which may have to be handled. Any attempt to normalise the large shells, now fabricated at a temperature above the critical range of about 900°C, would result in serious deformation because of the plasticity of the metal at that temperature. For metallurgical reasons, rapid cooling through the temperature range between 900° and 600°C would be necessary, but it would be impossible with hollow cylindrical vessels because of the danger of unequal stresses due to contraction and the considerable heat contained in the large mass of metal presented by the vessel.

The heat treatment to which vessels are subjected consists of bringing the metal to a temperature of 600°C, and maintaining it at that temperature for a period of 1 hours for each inch of thickness of the weld, the vessel being allowed to cool slowly afterwards in suitable screens. There can be no doubt that low-temperature stress relief treatment is beneficial. Suggestions have been made from time to time that a higher stress relief temperature ranging between 650° and 700°C should be employed. Whilst there is no doubt that the plasticity of the metal will increase with temperature, it has to be remembered that within this temperature range the carbide forms spheroidal aggregates at a relatively rapid rate, and any small benefit would only be gained at the expense of the structure of the metal.

Non-destructive Tests: In welded pressure vessels, test plates can easily be provided at each end of the longitudinal seam, these plates being tacked on prior to welding and butt-welded as part of the longitudinal seam of the vessel.

A welded cylinder with the test plates in position. After the welding is completed and the plates removed, they are given a controlled and recorded heat treatment exactly similar to that to which the drum is subjected, and are afterwards cut up into suitable specimens for transverse tensile, bend, all-weld-metal tensile, weld-metal density, and impact tests, whilst the material is also examined microscopically.

Such plates, however, cannot be produced when welding circumferential seams, as machine-welded circumferential scams are continuous. Some means, therefore, have to be found whereby it is possible to establish that the quality of the weld-metal shown by the mechanical tests of the test plates is identical with that in the whole length of the longitudinal and circumferential seams. Examination by X-rays has proved to be the most satisfactory means for attaining this end. By radiographic inspection of every inch of the weld and of the test plates, it is possible to establish a direct comparison between the weld-metal in the test plates and the weld-metal in the vessel itself.

The radiographic inspection is considerably simpler than might be supposed. The X-rays emitted by the tube penetrate the weld from the outside of the vessel, so that the radiograph is produced on a highly sensitised film carried in a cassette secured to the inner surface of the drum along the weld. The radiograph produced in a shadow picture, and the reading of these radiographs, whilst initially difficult, soon becomes with practice comparatively easy.

The most satisfactory way of obtaining a first-class knowledge of the meaning of radiographs is that of making welds under conditions which are deliberately varied, examining them by X-rays, making sections, and afterwards comparing these with the pictures produced and studying the history of the weld.

The code of the American Society of Mechanical Engineers provides a set of radiographs which are chosen to indicate acceptable or unacceptable welds, and this comparative basis for examination has proved satisfactory since its inclusion in the Code in 1931.

Acetylene Welding

Compared with even a few years ago, the standard of oxyacetylene welding, both in cost and quality, shows remarkable progress. The methods of making oxyacetylene welds can be classified under the following headings:

(1) Leftward welding.

(2) Rightward welding.

(3) Upward vertical welding (double reinforcement)

Method A: Welding from one sides only.

Method B: Welding simultaneously from both sides.

Upward Vertical Welding, Method "A." (Oxyacetylene Process.)

Direction of Welding: The welding rod, the flame and the seam are in the same vertical plane. The sheets to be joined are arranged in a vertical position, the weld being made in a vertical position.

Direction and Position of the Flame: The tip of the blowpipe, and consequently the flame, is inclined at an angle of 30° to the horizontal.

Direction and Position of the Welding Rod: The welding rod makes an angle of 20° with the horizontal, in a direction opposite to that of the tip of the blowpipe. The end of the rod is held in the molten metal and just within the edges to be joined.

Size of the Welding Rod: The diameter should be based upon the thickness of the sheet or plate and should be equal to one-half the thickness of the material to be welded.

Movement of the Flame: After forming a small hole by melting the two edges—the hole being maintained during the welding of the seam— only a steady upward movement in the direction of the seam is necessary.

Movement of the Welding Rod: The welding rod is drawn backward and forward in the molten metal.

As compared with the methods of rightward and leftward welding, it ensures complete control of the double reinforcement, as in rightward welding, but does not give the same reducing flame protection on the reverse side of the seam. It is, therefore, not so well adapted for metals which oxidise readily at a red heat. The power of the blowpipe being less, the speed of welding is consequently less, and the cost of welding slightly higher. The advantages include the decreased importance of the gap between the edges and the small demands made upon the concentration of the welder. Penetration is much more certain than with leftward welding.

Upward Vertical Welding, Method " B"

This method requires the services of two welders, welding simultaneously on either side of the seam.

Direction of Welding: The welding rod, the flame and the seam are in the same vertical plane. The plates to be joined are arranged in a vertical position, the welds being made in an upward direction.

Direction and Position of the Flames: The tip of each blowpipe, and consequently the flame, is inclined at an angle of 30° to the horizontal.

Direction and Position of the Welding Rods: Each welding rod makes an angle of 20° with the horizontal, and in a direction opposite to that of the tip of the blowpipe. The end of the welding rod is held in the molten metal, within the edges to be joined.

Size of the Welding Rods: The diameter of the welding rods should be 10 gauge.

Movement of the Flames: As soon as the first common pool of metal has been obtained, by each welder penetrating one-half of the plate at the same level, only a steady upward movement is necessary.

Movement of the Welding Rods: A transverse or zigzag movement is given to the end of the rod in the molten metal.

The advantages of this method are:

(1) Variations in the width of the gap between the edges do not seriously affect the results.

(2) The technique is rapidly acquired, and small demand is made on the concentration of the welders.

(3) First-class results are continuously obtained.

(4) Distortion is reduced to a minimum owing to the constant cross-section of the welded zone.

(5) Improved economy due to the low consumption of gases as a result of the better utilisation of the available heat.

(6) Within the recommended range of thicknesses no chamfering of the edges is necessary.

The soundness and quality of the welds obtained by this method make it particularly valuable where the parts have to support considerable stress, as in the case of welded pressure vessels, containers, tanks, etc., where the two sides are accessible. The protecting effect of the reducing-zone of the flame is obtained, an advantage not secured by leftward or upward vertical welding, method "A."

Welding Workmanship

With welded joints the allowable efficiency or ratio of joint-strength to the strength of the solid plate must depend upon a reasonably good standard of workmanship, since there can be 110 doubt that the quality of a welded joint is more easily affected by human inconsistencies than, say, a riveted joint. Defects which might be due to faulty workmanship are as follows:

(1) Imperfect fusion with parent-plate.

(2) Slag inclusions.

(3) Gas pockets.

(4) Contamination of weld-metal.

(5) Overheating of weld-metal or parent-plate.

Items (1), (2) and (3) and to some extent (4) will be detected by X-ray examination which must, in consequence, play an important part in the assessment of joint-efficiency. Item (4) indicates defects due to large fluctuations in arc-length, improper manipulation of electrodes, or, in the case of oxyacetylene welding, an improper admixture of the gases in the flame. Under a careful system of supervision it is unlikely that these defects will be extensive throughout a welded seam, and the total effect of such local contamination (local not only in relation to the length of the seam but also in relation to the depth) need not be regarded with undue alarm. With regard to (5), serious overheating (that is to say, overheating which cannot be rectified by subsequent heat treatment) is unusual in the case of electric arc welding as carried out in high-class welding shops, because the factors of current values and electrode fluidity should always be subject to special control.

A possible cause of overheating, however, would be the maintenance of the arc in one position for too long a period. Fortunately, the defect would be readily revealed by visual inspection, due to the fact that it would be accompanied by serious undercutting at that particular position.

The purport of the foregoing remarks on workmanship is to show that this factor can, without difficulty, be accounted for in assessing the joint-efficiency, by providing:

(1) Adequate supervision of the welding work,

(2) Careful inspection of each run of weld-metal,

(3) X-ray examination of the finished joint.

It is the practice with reputable firms in Great Britain for an experienced supervisor to be responsible for the detailed inspection of each run of weld-metal as deposited in the welded joints of important pressure vessels. His duties also include the supervision of the arrangement of the vessel for welding, and the control of the current values.

Such close supervision must be regarded as an essential factor in fusion-welding, as it is only by means of this supervision that the reliability of workmanship can be accepted as equivalent to that involved in high-class riveting for boiler construction.

Resources for Welded Pressure Vessels

In Great Britain, for steel pressure vessels the material usually specified is mild steel made by the acid or basic open-hearth process. The ultimate tensile strength of the plates lies between the limits of 26 and 30 tons per sq. in., or between 28 and 32 tons per sq. in., depending upon the tensile range specified. The minimum elongation required on a gauge-length of 8 ins. is 23 per cent, for the 26-30 ton steel and 20 per cent, in the case of 28 32 ton steel. In certain cases, the plates are normalised after rolling, and it is good practice to anneal (stress-relieve) the plates after bending them to the required shape.

Lower-tensile steel plates would be accompanied by lower carbon and manganese contents, but the range of silicon indicated in the above analysis permits the use of steel in either the fully " killed " or "semi-killed" condition.

It has long been the practice in riveted-boiler construction to use steel which may be said to be fully " killed," and which in consequence possesses a silicon content of about 0.2 per cent. For the purposes of welding, however, opinion turning in favour of low-silicon steel.

The requirements of other authorities in regard to plate-material for welding are as follows:

United States: The regulations here are based on the requirements of the Power Boiler Construction Code of the American Society of Mechanical Engineers. These requirements permit a range of steel for welding purposes; the maximum allowable carbon content for steel considered to be of weldable quality is 0.35 per cent.

Germany: In Germany regulations are in force in respect of forge welding of boiler-drums, but in general, fusion welding is only permitted in cases where the welded seam is covered by butt-straps. It is the practice, however, to give special permission, by means of Ministerial Decrees, to certain firms to enable them to manufacture fusion-welded boiler-drums without straps. The existing regulations are now being revised and a draft of new regulations has recently been submitted by the Committee appointed for that purpose.

The analysis of plate material most commonly used in Germany for welding purposes is:

Carbon...... from 0.1 to 0.3 per cent.

Manganese...... 0.5 per cent, (max.)

Phosphorus...... 0.05 „ „ „

Sulphur...... 0.05 „

Silicon...... up to 0.25 per cent.

Steel associated with this analysis is usually supplied in two ranges:

1. From 34 to 41 kgs. per sq. mm., with an elongation of from 28 to 25 per cent, on 200 mm., and a reduction of area of from 80 to 70 per cent.
2. From 41 to 48 kgs. per sq. mm., with an elongation of from 25 to 20 per cent, on 200 mm., and a reduction of area of from 70 to 60 per cent.

Until recently steel of 48 kgs. per sq. mm. was considered to be the limit for good weldability.

Whilst the existing German regulations do not include any stipulations in regard to impact-testing of boilerplate, the German Steam Boiler Committee have given considerable thought to the subject, and it is of interest to note that a Charpy test value of 10 metre-kgs. per sq. cm. is regarded as appropriate for plate thicknesses up to 15 mm. and 8 metre-kgs. per sq. cm. for plate thicknesses above 15 mm. These values are quoted in connection with plate material of the 34r-41 kgs. per sq. mm. quality, but in view of the well-known lack of consistency in impact-test results on rolled-steel plates the figures should not be regarded as a final criterion.

Recent progress in Germany has, however, made it possible to adopt welding also for steels of higher tensile strength, and for slightly alloyed steels. These steels include the above-mentioned qualities of mild steel with the addition of 0.5 per cent, of molybdenum, the steel having a tensile strength of from 35 to 50 kgs. per sq. mm.; ordinary mild steel containing 0.25 per cent, of molybdenum and 0.25 per cent, of copper, with a tensile strength of from 41 to 53 kgs. per sq. mm.; a special non-ageing carbon steel with a tensile strength of from 47 to 56 kgs. per sq. mm.; and this steel with the addition of 0.25 per cent, of molybdenum and 0.25 per cent, of copper.

The above-mentioned steels are approved for electric-arc fusion welding, while in the case of water-gas welding, approval has been given for the use of a mild steel containing 0.25 per cent, of molybdenum and 0.25 per cent, of copper, with a tensile strength of from 44 to 56 kgs. per sq. mm., and for a steel having reduced ageing properties and a tensile strength of from 44 to 53 kgs. per sq. mm.

Switzerland: In Switzerland the regulations governing the construction of welded boiler-drums are issued by the Swiss Association of Owners of Steam Boilers (Schweizerischer Verein von Dampfkessel-Besitzern). These regulations, promulgated in January 1932, specify two qualities of mild-steel plate, and whilst both qualities may be used for welding.

Sweden: In Sweden, whilst considerable development has taken place in regard to the welding of structures and unfired pressure vessels, the welding of boiler-drums is not yet an accepted practice. No detailed regulations have been issued in respect of welded pressure vessels, but in general it can be said that the plate-material used for water-gas welding has a minimum tensile strength of 38 kgs. per sq. mm. and a carbon content of from 0.12 to 0.15 per cent.

The material is of the low-silicon type. For fusion welding, higher tensile strength is generally used, the material having a range of from 44 to 55 kgs. per sq. mm. and a carbon content of from 0.16 to 0.20 per cent.

Italy: In Italy, there are no detailed regulations except for a Ministerial Decree which deals broadly with the application of welding to pressure vessels, and which merely states that the plates are to be of a quality suitable for welding.

It is, perhaps, remarkable that although evidence indicates that Continental engineers are fully abreast of welding developments, no authoritative attempt has been made to establish a construction code governing the details of welding as applied to pressure vessels.

Even today the only authority to issue such a code in Europe is Lloyd's Register of Shipping, which, in July 1934, published " Tentative Requirements for Fusion Welded Pressure Vessels intended for Land Purposes."

Certain European authorities contemplate the acceptance of welded pressure vessels in their official regulations in that they do riot prohibit them. Others, such as the Swiss Association of Owners of Steam Boilers, go so far as to legislate for the design of welded pressure vessels, leaving the details of construction and testing to the discretion of the expert inspectors. In other cases, for example in Czechoslovakia, construction is generally carried out in accordance with the requirements of the Boiler Construction Code of the American Society of Mechanical Engineers.

Rules for Design of Welded Joints of Pressure Vessels

Joint Efficiencies: In considering the question "What is a reasonable figure to assign for welded joint-efficiency"? it would be well to take note of the joint-efficiencies and factors of safety allowed by those authorities which have legislated for welded joints in pressure vessels.

Great Britain: The only published rules governing design of Class I fusion-welded pressure, vessels are Lloyd's Register's "Tentative Requirements for Fusion Welded Pressure Vessels intended for Land Purposes." These are at present under revision, but in the meantime may be regarded as the accepted standard for both land and marine work in Great Britain.

United States of America: The Power Boiler Construction Code of the American Society of Mechanical Engineers permits a joint-efficiency of 90 per cent., combined with a factor of safety of 5. The formula for working pressure in lbs. per sq. in. is TS × *t x* E ~FS × R

where TS = the ultimate tensile strength of the plate in lbs./sq. in.

t = the minimum thickness of the shell-plate in ins.

E = the joint efficiency = 90 per cent.

FS — the factor of safety = 5.

R = the internal radius of the shell in ins. If the shell thickness is greater than 10 per cent, of the radius, the outer radius is to be used for R.

The specification governing the welding of pressure vessels issued by the Bureau of Engineering of the US Navy, states that the joint-efficiency of welded joints under the cognisance of this Bureau shall be taken as 80 per cent, of the strength of the parent-metal, except where the quality of the weld can be and is fully explored by radiographic or exographic photography, or other method satisfactory to the Bureau, in which case a joint-efficiency of 90 per cent, will be acceptable.

The joint "Code for the Design, Construction, Inspection, and Repair of Unfired Pressure Vessels for Petroleum Liquids and Gases", issued by the American Petroleum Institute in conjunction with the American Society of Mechanical Engineers Schuster (*Proc. Inst. Mech. Eng.,* 1930, 372) suggested Provisional Rules for Fusion-Welded Non-Fired Pressure Vessels, and gave the following recommendations about the method of fabricating pressure vessels:

1. *Scope of Rules:* The rules apply to vessels containing air or non-corrosive gas at a temperature not exceeding 600°F, and more especially when the diameter does not exceed 36 ins. For vessels with a diameter not exceeding

20 ins., the rules do not apply when the working pressure is 30 lbs. per sq. in. or less.

2. *Metfiod of Welding:* The rules are applicable to vessels welded by the metallic arc process with a flux-coated electrode or by the oxyacetylene process.

   Note 1: The rule excludes carbon-are-welding, but in. special instances, where the design is approved, the carbon-arc process is permissible for the welding of flanges to tubes, and for the welding of dished ends to shell plates that are secured independently of the weld.

   Note, 2: It is recommended, especially when, the welding has been carried out by the carbon-arc or acetylene process, that fusion-welded parts should be normalised in a muffle furnace. The temperature should be maintained for a quarter of an hour after the article has been heated uniformly.

   As an alternative, though less effective, where serious damage from scaling or warping would result if it were normalised, the article should be heated uniformly to a temperature of 0-50°C, and then allowed to cool.

3. *Strength of Joint:* The rated strength of a welded joint shall at no time exceed half the strength determined by tests on reasonably well-made joints, made to similar form and subjected to similar stresses.

4. *Material:* The plate shall be of boiler-quality mild steel with a nominal breaking strength of 26 tons per sq. in. The strength shall lie between the limits of 24 to 28 tons per sq. in., but it shall be assumed that the strength of the plate when welded does not exceed 24 tons per sq. in.

5. *Plate Thickness:* The plate of a welded vessel exceeding 12 ins. in diameter.

6. *Factor of Safety:* The minimum factor of safety allowable on the rated strength of a welded joint shall be 4.
7. *Hydrostatic Test:*
   (a) The vessel shall be subjected to a hydrostatic test at not less than twice the working pressure, during which it shall be subjected to a hammer test; the pressure shall then be reduced by more than 50 per cent, of its value, after which it shall be again raised to the original value for 3 minutes.

   (b) The hammer test shall consist of the plate on both sides of all welds being struck sharp vibratory blows with a hammer not less than 2 lbs. in weight, the blows being struck 2 to 3 ins. apart and close up to the seam. The blows shall be applied as rapidly and heavily as possible, consistent with the metal not being indented or distorted.

   (c) When the factor of safety of the vessel exceeds 4, the test pressure shall be 50 per cent, of the bursting pressure, calculated on the rated strength of the welded joints.
8. *Stresses Imparted by Flexible Ends:* Where a joint or a portion of a joint, circular or longitudinal, is subjected to a bending action owing to the flexibility of an end, it shall be deemed that the strength of the joint is reduced by 10 per cent.

   This shall not apply to a longitudinal seam of which the ends are reinforced in an approved manner to resist bending stresses.
9. *Lap-Welded Joints:*
   (a) For a longitudinal seam, a lap weld with a single fillet shall at no time be permitted, and a lap weld with a double fillet shall only be permissible.

(b) For a circular seam or for affixing an end plate when the shell plate is not constricted, i.e. turned in over the flange of the end plate, a lap-welded joint with a single fillet shall only be permitted.

(c) When a lap-welded joint has a double fillet the internal fillet shall be accessible for inspection. Where there is no ready means for inspection it shall be assumed that the fillet is non-existent, and accordingly for a circumferential seam the joint shall be rated as if it had a single fillet only and for a longitudinal scam it shall be prohibited.

10. *Fillet Welds:*

(a) With a fillet of a lap-welded joint the surface of contact in tension shall extend over the full thickness of the plate.

(b) A fillet of a lap-welded joint shall by preference have a convex profile. At no time shall the thickness at the throat be less than 0.69 times the plate thickness t.

Note: The throat thickness is defined as the minimum depth of the weld. Where the two contact surfaces of a fillet are equal, the rule requires that the contour shall show no appreciable concavity.

(c) With a fillet weld affixing a fitting to the wall of a vessel or a pipe, the throat shall have a minimum thickness of 0. 58Å, where h is the minimum height allowable for the surface of contact in shear.

Note 1: Where the two surfaces of contact of the fillet are equal, this throat thickness is attained by a fillet.

Note 2: The value required for the dimension h depends on the details of the design and manufacture of the joint.

11. *Butt-Welded Joints:*

(a) When the diameter of the vessel does not exceed 24 ins., the strength of a butt-joint with the plate welded at only one side shall be deemed to be reduced to half the strength of a butt-joint with the plate welded at both sides. For larger diameters this form of construction shall at no time be permitted.

(b) When the inside surface of a butt-joint is not accessible for inspection, it shall be assumed that the plate is welded at one side only.

(c) The edges of a plate jointed by a butt-weld shall be bevelled, and the included angle of the weld-metal shall be not less than 90° for either a single or a double V weld.

12. *Design and Attachment of Dished Kinds:*

(a) A dished end plate shall have an inside radius not greater than the internal diameter of the shell. The thickness of metal shall be such that the working pressure or the allowable stress to comply with rule 13 is not exceeded, but in no instance shall it be less than the thickness of the shell, and for vessels exceeding 12 ins. in diameter the minimum thickness shall be 1 in.

(b) When a dished end is flanged, the length of the flange shall be not less than 1 in. for an end convex to the pressure, nor less than 2 in. with the end concave. The minimum length of the flange shall at no time be less than four times the thickness of the plate. An increase above these specified minimum values is desirable.

3

Chemistry's Application in Agriculture

Agrichemical (or agrochemical), a contraction of agricultural chemical, is a generic term for the various chemical products used in agriculture. In most cases, *agrichemical* refers to the broad range of insecticides, herbicides, and fungicides, but it may also include synthetic fertilizers, hormones and other chemical growth agents, and concentrated stores of raw animal manure.

Most agrichemicals are toxic, and all agrichemicals in bulk storage pose significant environmental and/or health risks, particularly in the event of accidental spills. In many countries, access to and use of agrichemicals is highly regulated. Government-issued permits for purchase and use of approved agrichemicals may be required, and significant penalties can result from misuse, including improper storage resulting in spillage. On farms, proper storage facilities and labelling, emergency clean-up equipment and procedures, and safety equipment and procedures for handling, application and disposal are specific areas of concern, often subject to mandatory standards and regulations. Usually, the regulations are carried out through the registration process.

Agricultural chemistry is the science of the causes and effects of biochemical reactions related to plant and animal growth. Agricultural Chemists study life processes for obtaining food, Agricultural chemistry focuses on chemical compositions and changes involved in the production, protection, and use of crops and livestock.

It seeks to control and understand the processes by which humans obtain food and fibre for themselves and feed for their animals. Agricultural chemists work with food producers to increase yields, improve quality, and reduce costs. They also study the causes and effects of biochemical reactions related to plant and animal growth, seek ways to control these reactions, and develop chemical products that provide help in controlling these reactions. Chemical products developed to assist in the production of food, feed, and fibre include herbicides, fungicides, insecticides, plant growth regulators, fertilizers, and animal feed supplements.

Regulate the Safe use of Chemicals: Through research and experimentation, agricultural chemists increase our understanding of the impact of chemicals in the food stream and advise farmers, the government, and the agricultural industry on the effective and safe use of chemicals to improve and expand production. Every year in the United States, millions of pounds of chemicals are used in herbicides, fungicides, and insecticides that are applied to crops.

Some industrial scientists in this field use their expertise in agriculture and chemical processes to make sure these products are safe for humans and for the environment. These chemicals are applied in huge quantities to commercial crops. In test fields, the air, plants, soil, and water are tested for the presence of these chemicals or their metabolites—compounds which result from the breakdown of the chemicals. A significant presence may be dangerous to animals and humans.

Other agricultural chemists do similar work for government agencies such as the US Department of Agriculture (USDA). These agricultural scientists focus on the environmental fate of pesticides, finding out where they go, which can sometimes be unexpected. Chemists test for these chemicals in the atmosphere, surface water, and groundwater. Agricultural chemists also work at the Environmental Protection Agency (EPA) and the Food and Drug Administration (FDA), the US government regulatory agencies for the industry. In some cases, the work that both industry and government do results in changes in the way that some chemicals are used.

Have a Variety of Opportunities: Some chemists develop molecules used in herbicides or other pesticides, and others develop molecules and new compounds that effectively reduce damage to crops from pests and disease but are less toxic to animals and humans who eat the food. Still others help to improve the cost-effectiveness of food production or are focused on other business-related aspects of the field.

Agricultural biotechnology is the fastest growing sub-discipline within agricultural chemistry and is currently focused in three major areas: genetic engineering of crops to be more herbicide-tolerant or less apt to be killed along with the weeds during herbicide treatment; genetic engineering of produce, to improve taste and colour and promote longer shelf life; and the improvement of plants' natural tolerance to certain pests. An example of the latter is the Calgene Corporation's work on introducing the Bacillus thuringiensis (Bt) protein into its proprietary cotton varieties.

This causes the plants to produce a Bt toxin that kills Heliothis, a principal cotton insect pest. Crops engineered to be herbicide tolerant are thought to be safer and more environmentally friendly than other crops.

Process development scientists at agricultural firms seek ways to make their products efficiently. They work with the research and development departments to identify market needs and find ways to meet them. These scientists concern themselves most with how to produce needed chemicals at a reasonable cost to their buyers and with minimal environmental impact.

Work with Other Scientists: Agricultural chemists work a lot with scientists in other fields, including agronomists, biologists, toxicologists, and biochemists. There is a wide spectrum of research projects in this field. Thus, it is important to be able to work, or at least to be conversant, in other fields. An agricultural chemist should be as well-rounded as possible. Although the work is very team-based, there's still autonomy in research as long as the research ultimately has an application that will benefit agriculture.

Agricultural chemistry is not a distinct discipline. It ties together with genetics, physiology, microbiology, entomology, and other sciences that contribute chemical techniques to agriculture. For example, agricultural chemists help develop more productive plant and animal strains, determine the kinds and amounts of nutrients needed for optimum growth of plants and animals, and determine a soil's ability to provide essential nutrients for the support of crops or livestock. Every scientific discipline that contributes to agricultural progress depends in some way on chemistry.

Formation of Diammonium Phosphate

Diammonium phosphate (DAP) (chemical formula $(NH_4)_2HPO_4$ is one of a series of water-soluble ammonium phosphate salts which can be produced when ammonia reacts with phosphoric acid.

DAP is used as a fertilizer and a fire retardant. When applied as plant food, it temporarily increases the soil pH (more basic), but over a long term the treated ground becomes more acidic than before upon nitrification of the ammonium. It is incompatible with alkaline chemicals because its ammonium ion is more likely to convert to ammonia in a high-pH environment.

DAP is sometimes used as a yeast nutrient for brewing mead, and is an additive in some brands of cigarettes.

Properties of Potassium Chloride

The chemical compound potassium chloride (KCl) is a metal halide composed of potassium and chlorine. In its pure state it is odourless. It has a white or colourless vitreous crystal, with a crystal structure that cleaves easily in three directions. Potassium chloride crystals are either simple cubic or face-centred cubic depending on what atoms are involved. If only potassium or chlorine atoms are considered, then the structured is face-centred cubic. However, both atoms form a crystal with a simple cubic structure:

> X-ray diffraction analysis will yield a simple cubic structure. Potassium chloride is also commonly known as "Muriate of Potash". Potash varies in colour from pink or red to white depending on the mining and recovery process used. White potash, sometimes referred to as soluble potash, is usually higher in analysis and is used primarily for making liquid starter fertilizers. KCl is used in medicine, scientific applications, food processing and in judicial execution through lethal injection. It occurs naturally as the mineral sylvite and in combination with sodium chloride as sylvinite.

Formation of Potassium Chloride

Potassium chloride can react as a source of chloride ion. As with any other soluble ionic chloride, it will precipitate insoluble chloride salts when added to a solution of an appropriate metal ion:

$$KCl(aq) + AgNO_3(aq) \longrightarrow AgCl(s) + KNO_3(aq)$$

Although potassium is more electropositive than sodium, KCl can be reduced to the metal by reaction with metallic sodium at 850°C because the potassium is removed by distillation:

$$KCl(l) + Na(l) \longrightarrow NaCl(l) + K(g)$$

This method is the main method for producing metallic potassium. Electrolysis (used for sodium) fails because of the high solubility of potassium in molten KCl. As with other compounds containing potassium, KCl in powdered form gives a lilac flame test result.

Manufacture/Extraction

Potassium chloride occurs naturally as sylvite, and it can be extracted from sylvinite. It is also extracted from salt water and can be manufactured by crystallisation from solution, flotation or electrostatic separation from suitable minerals. It is a by-product of the making of nitric acid from potassium nitrate and hydrochloric acid.

Uses

The majority of the potassium chloride produced is used for making fertilizer, since the growth of many plants is limited by their potassium intake. As a chemical feedstock it is used for the manufacture of potassium hydroxide and potassium metal.

It is also used in medicine, scientific applications, food processing, as a sodium-free substitute for table salt (sodium chloride), and in judicial execution through lethal injection. It is sometimes used in water as a completion fluid in oil and gas operations. KCl is useful as a beta radiation source for calibration of radiation monitoring equipment because natural potassium contains 0.0118 per cent of the isotope

K^{40}. One kilogram of KCl yields 16350 Becquerel of radiation consisting of 89.28 per cent beta and 10.72 per cent gamma with 1.46083 MeV.

Biological/Medical Properties

Potassium is vital in the human body and oral potassium chloride is the common means to replenish it, although it can also be diluted and given intravenously. It can be used as a salt substitute for food, but due to its weak, bitter, unsalty flavour, it is usually mixed with regular salt, sodium chloride, for this purpose to improve the taste.

Medically it is used in the treatment of hypokalaemia and associated conditions, for digitalis poisoning, and as an electrolyte replenisher. Brand names include K-Dur, Klor-Con, Micro-K, and Kaon Cl. Side effects can include gastrointestinal discomfort including nausea and vomiting, diarrhoea and bleeding of the gut. Overdoses cause hyperkalemia which can lead to paraesthesia, cardiac conduction blocks, fibrillation and arrhythmias, also sclerotic.

Solubility of KCl in various solvents
(g KCl / 100 g of solvent at 25°C)

H_2O	36
Liquid ammonia	0.04
Liquid sulphur dioxide	0.041
Methanol	0.53
Formic acid	19.2
Sulpholane	0.004
Acetonitrile	0.0024
Acetone	0.000091
Formamide	6.2
Acetamide	2.45
Dimethylformamide	0.017 - 0.05

Physical Properties

Potassium chloride has a crystalline structure like many other salts. Structure: face-centred cubic. Lattice Constant: roughly 6.3 angstroms.

In chemistry and physics it is a very commonly used as a standard, for example as a calibration standard solution in measuring electrical conductivity of (ionic) solutions, since carefully prepared KCl solutions have well-reproducible and well-repeatable measurable properties.

Precautions

Orally it is toxic in excess; the LD_{50} is around 2500 mg/kg (meaning that a person weighing 75 kg (165 lb) would have to consume about 190 g (6.7 oz); regular salt is about as toxic).

Intravenously this is reduced to just over 100 mg/kg but of more concern are its severe effects on cardiac muscles; high doses can cause cardiac arrest and rapid death. Indeed, a massive overdose of intravenous potassium chloride is used to stop the heart in execution by lethal injection.

Properties of Potassium Nitrate

The chemical compound potassium nitrate is a naturally occurring mineral source of nitrogen, but makes up the critical oxidising component of gun powder, was necessary for burning fuse technologies including slow matches and since it readily precipitates, was widely "harvested" since the Late Middle Ages and Early Modern era through the 19th century from urine from which it was forced to crystallise in various odiferous ways. It is a nitrate with chemical formula KNO_3.

Its common names include saltpetre (from Medieval Latin sal petrae: "stone salt" or possibly "Salt of Petra"), American English saltpetre, Nitrate of potash and nitre. The name Chile saltpetre is also applied to sodium nitrate, which while related to explosives as well, is a very different compound.

Potassium nitrate is the oxidising component of black powder. Prior to the large-scale industrial fixation of nitrogen through the Haber process, a major source of Potassium nitrate was the deposits crystallising from cave walls or the drainings of decomposing organic material. Dung-heaps were a particularly common source: ammonia from the decomposition of urea and other nitrogenous materials would undergo bacterial oxidation to produce nitrate.

It was and is also used as a component in some fertilizers. When used by itself as a fertilizer, it has an NPK rating of 13-0-44 (indicating 13 per cent, 0 per cent, and 44 per cent of nitrogen, phosphorus, and potassium, by mass, respectively). Potassium nitrate was once thought to induce impotence, and was rumoured to be added to institutional food as an anaphrodisiac. If true, the treatment was ineffective, as potassium nitrate has no such properties.

Manufacture

Historically, nitre-beds were prepared by mixing manure with either mortar or wood ashes, common earth and organic materials such as straw to give porosity to a compost pile typically 1.5 metres high by 2 metres wide by 5 metres long. The heap was usually under a cover from the rain, kept moist with urine, turned often to accelerate the decomposition and leached with water after approximately one year. The liquid containing various nitrates was then converted with wood ashes to potassium nitrates, crystallised and refined for use in gunpowder.

In more rural times, urine was collected and used in the manufacture of gunpowder. Stale urine was filtered through a barrel full of straw and allowed to continue to sour for a year or more. After this period of time, water was used to wash the resulting chemical salts from the straw.

This slurry was filtered through wood ashes and allowed to dry in the sun. Saltpetre crystals were then collected and added to brimstone and charcoal to create black powder. Today, most potassium nitrate comes from the vast deposits of sodium nitrate ($NaNO_3$, nitratine) in the Chilean deserts. The sodium nitrate is purified and then reacted in solution with potassium chloride (KCl, sylvite), from which the less-soluble potassium nitrate is precipitated out.

Applications

One of the most useful applications of potassium nitrate is in the production of nitric acid, by adding concentrated sulphuric acid to an aqueous solution of potassium nitrate, yielding nitric acid and potassium hydrogen sulphate which are separated through fractional distillation.

Potassium nitrate is also used as a fertilizer, in model rocket propellant, and in several fireworks such as smoke bombs, in which a mixture with sugar produces a smoke cloud of 600 times their own volume.

The ratio for smoke bombs using sucrose (powdered sugar) and potassium nitrate is $40(C_{12}H_{22}O_{11}):60(KNO_3)$. It can be used as is, or carefully melted together using a hot plate.

In the process of food preservation, potassium nitrate has been a common ingredient of salted meat, but its use has been deprecated. In the European Union, it is referred to as E252.

Potassium nitrate is also the main component (usually about 98 per cent) of tree stump remover; it accelerates the natural decomposition of the stump.

It has also been used in the manufacture of ice cream and can be found in some toothpastes for sensitive teeth. Recently, the use of potassium nitrate in toothpastes 1for sensitive teeth has increased dramatically, despite the fact that it has not been conclusively shown to help dental hypersensitivity.

Potassium nitrate is also one of the three parts of gunpowder, along with powdered charcoal (substantially carbon) and sulphur. It acts as an oxidiser, oxidising the carbon and sulphur to their respective oxides.

Properties of Potassium Sulphate

Potassium sulphate (K_2SO_4) is a non-flammable white crystalline salt which is soluble in water. The chemical is commonly used in fertilizers, providing both potassium and sulphur.

Potassium sulphate (K_2SO_4) has been known since early in the 14th century, and it was studied by Glauber, Boyle and Tachenius. In the 17th century it was named *arcanuni* or *sal duplicatum*, as it was a combination of an acid salt with an alkaline salt.

Natural Resources

Natural resources of potassium sulphate are minerals abundant in the Stassfurt salt. These are cocrystalisations of potassium sulphate and sulphates of magnesium calcium and sodium. The minerals are

- Kainite $MgSO_4 \cdot KCl \cdot H_2O$
- Schonite $K_2SO_4 \cdot MgSO_4 \cdot 6\ H_2O$
- Leonite $K_2SO_4 \cdot MgSO_4 \cdot 4\ H_2O$
- Langbeinite $K_2SO_4 \cdot 2\ MgSO_4$
- Glaserite $K_3Na(SO_4)_2$
- Polyhalite $K_2SO_4 \cdot MgSO_4 \cdot 2\ CaSO_4 \cdot 2\ H_2O$

From some of the minerals like kainite, the potassium sulphate can be separated, because the corresponding salt is less soluble in water.

With potassium chloride kieserite $MgSO_4 \cdot 2\ H_2O$ can be transformed and then the potassium sulphate can be dissolved in water.

Manufacture

- Corresponding to the first synthesis form Johann Glauber potassium sulphate can be synthesised by the decomposition of potassium chloride with sulphuric acid, hydrogen chloride evaporates and can be used to produce hydrochloric acid.

- The Hargreaves method is basically the same process with different starting materials. Sulphur dioxide, oxygen and water (the starting materials for sulphuric acid) are reacted with potassium chloride. Hydrochloric acid evaporates off.
- It is obtained as a by-product in many chemical reactions including the production of nitric acid. This can be done by mixing the following: 2 Parts Potassium Nitrate to 1 Part Sulphuric Acid (both by weight).

$$2KNO_3 + H_2SO_4 \longrightarrow 2HNO_3 + K_2SO_4$$

To purify the crude product, it can be dissolved in hot water and then filtered and cooled, causing the bulk of the dissolved salt to crystallise with characteristic promptitude.

Properties

The anhydrous crystals form a double six-sided pyramid, but are in fact classified as rhombic. They are transparent, very hard and have a bitter, salty taste. The salt is soluble in water, but insoluble in solutions of potassium hydroxide, or in absolute ethanol. It melts at 1078°C.

Uses

The principal use of potassium sulphate is as a fertilizer. The crude salt is also used occasionally in the manufacture of glass.

Potassium Hydrogen Sulphate

Potassium hydrogen sulphate or bisulphate, $KHSO_4$, is readily produced by mixing K_2SO_4 with an equivalent no. of moles of sulphuric acid. It forms rhombic pyramids, which melt at 197°C. It dissolves in three parts of water of 0°C. The solution behaves much as if its two congeners, K_2SO_4 and H_2SO_4, were present side by side of each other uncombined; an excess of ethanol the precipitates normal sulphate (with little bisulphate) with excess acid remaining.

Similar is the behaviour of the fused dry salt when heated to several hundred degrees; it acts on silicates, titanates, etc., the same way as sulphuric acid that is heated beyond its natural boiling point does. Hence it is frequently used in analytical chemistry as a disintegrating agent.

Prill

A prill is a small aggregate of a material, most often a dry pellet, usually formed from a liquid. Prills are formed by allowing drops of the prill substance to congeal in mid-air by being dripped from the top of a tall prilling tower. Fertilizers (ammonium nitrate, urea, NPK fertilizer) and some detergent powders are commonly manufactured as prills.

Superphosphate: Brief Summary

Superphosphate is a fertilizer produced by the action of concentrated sulphuric acid on powdered phosphate rock.

$$Ca_3(PO_4)_2 + 2H_2SO_4 \longrightarrow Ca(H_2PO_4)_2 + 2CaSO_4$$

History and Development

In the 1840s, scientists found that coprolites could be dissolved in sulphuric acid to produce what became known as superphosphate.

In 1840, Justus Von Liebig wrote, "The crops on the field diminish or increase in exact proportion to the diminution or increase of the mineral substances conveyed to it in manure." Von Liebig was the first to discover that phosphate of lime in bone meal could be rendered more readily available to plants by treatment with sulphuric acid. Sir John Bennett Lawes about the same time discovered that phosphate rock underwent the same reaction and could be used as a source ingredient.

Bennett Lawes was the first to manufacture superphosphate at his factory in Deptford, England in 1842.

A large market for superphosphate was created in the second half of the 20th century by the development of aerial topdressing in New Zealand which allowed superphosphate to be spread economically over large areas.

Superphosphate can be created naturally in large quantities by the action of guano, or bird faeces, resulting in deposits around sea bird colonies which can be mined. The most famous mining site is the island of Nauru in the South Pacific much of the "soil" from which was mined, creating temporary wealth for the inhabitants, but destroying their environment.

Uses of Triple Superphosphate

Triple Superphosphate is a fertilizer produced by the action of concentrated phosphoric acid on ground phosphate rock.

The active ingredient of the product, monocalcium phosphate, is identical to that of superphosphate, but without the presence of calcium sulphate that is formed if sulphuric acid is used instead of phosphoric acid. The phosphorus content of triple superphosphate (17 - 23 per cent P; 44 to 52 per cent P_2O_5) is therefore greater than that of superphosphate (7 - 9.5 per cent P; 16 to 22 per cent P_2O_5). Triple superphosphate was the most common phosphate (P) fertilizer in the USA until the 1960s, when ammonium phosphates became more popular. It is produced in granular and non-granular form and is used both in fertilizer blends (with potassium and nitrogen fertilizers) and by itself.

UAN

UAN is a solution of urea and ammonium nitrate in water used as a fertilizer. The combination of urea and ammonium nitrate has an extremely low critical relative humidity (18 per cent at 30°C) and can therefore only be used in liquid fertilizers. The most commonly used grade of these fertilizer solutions is UAN 32.0.0 (32 per cent N), which consists of 45 per cent ammonium nitrate, 35 per cent urea and only 20 per cent water. Other grades are UAN 28, UAN 30 and UAN 18. The solutions are quite corrosive towards mild steel (up to 500 MPY on C1010 steel) and are therefore generally equipped with a corrosion inhibitor to protect tanks, pipelines, nozzles, etc.

Physical and Chemical Characteristics of Urea Ammonium Nitrate Solutions

The solutions contain a remarkably low amount of water and nevertheless have a low salt-out temperature:

Physical and Chemical Characteristics of UAN Solutions

Grade, % N:	28	30	32
Ammonium Nitrate (%):	40	42	45
Urea (%):	30	33	35
Water (%):	30	25	20
Specific Gravity at 16°C:	1.283	1.303	1.320
Salt-out Temperature (°C):	-18	-10	-2

Formation of Agricultural Lime

Agricultural lime is a soil additive made from pulverised limestone or chalk. The primary active component is calcium carbonate. Additional chemicals vary depending on the mineral source and may include calcium oxide, magnesium oxide and magnesium carbonate.

Effects on soil:

- It increases the pH of acidic soils.
- It provides a source of calcium for plants.
- It permits improved water penetration for acidic soils.

Lime may occur naturally in some soils but may require addition of sulphuric acid for its agricultural benefits to be realised. Gypsum is also used to supply calcium for plant nutrition but will not raise the soil's pH. In livestock farming, hydrated lime can be used, as a 'disinfectant' measure, producing a dry and alkaline environment, in which bacteria do not readily multiply.

Spinner style lime spreaders are generally used to spread agricultural lime on fields. Several companies such as Stoltzfus Spreaders manufacture spreaders for this purpose.

Formation of Ammonium Sulphate

Ammonium sulphate, $[NH_4]_2[SO_4]$ sometimes, is an inorganic chemical compound commonly used as a fertilizer. It contains 21 per cent nitrogen as ammonia and 24 per cent sulphur as sulphate. Its molecular formula is closely related to Mohr's salt an ammonium sulphate analogue with an iron atom within the compound.

Properties

Ammonium sulphate is not soluble in alcohol or liquid ammonia, but is highly hygroscopic and spontaneously absorbs water from the air at relative humidity > 81 per cent.

Synthesis

Ammonium sulphate is prepared commercially by reacting ammonia with sulphuric acid (H_2SO_4). Ammonium sulphate is prepared commercially from the ammoniacal liquor of gasworks and is purified by recrystallisation. It forms large rhombic prisms, has a somewhat saline taste and is easily soluble in water. The aqueous solution on boiling loses some ammonia and forms an acid sulphate.

Uses

It is used largely as an artificial fertilizer for alkaline soils. In the soil the sulphate ion is released and forms sulphuric acid, lowering the pH balance of the soil (as do other sulphate compounds such as aluminium sulphate), while contributing essential nitrogen for plant growth.

It is also used as an agricultural spray adjuvant for water soluble insecticides, herbicides, and fungicides. There it functions to bind iron and calcium cations that are present in both well water and plant cells. It is particularly effective as an adjuvant for 2,4-D (amine), glyphosate, and glufosinate herbicides.

It is also used in the preparation of other ammonium salts. In biochemistry ammonium sulphate precipitation is a common method for purifying proteins.

Uses of Auxigro

Auxigro is a chemical-based growth-enhancer that is currently approved by the EPA for spray on fruits, vegetables, and grains. Its use is controversial because it contains glutamic acid, which is believed to cause undesirable illnesses in people sensitive to it.

Calcium Dihydrogen Phosphate

Calcium dihydrogen phosphate (also called mono-calcium orthophosphate) $Ca(H_2PO_4)_2$ is a chemical compound.

Decomposes at 203°C. The common $Ca(H_2PO_4)_2 \cdot H_2O$ releases a water molecule before it melts at 109°C.

Fertilizer

Phosphorus is an important nutrient and so is a common component of fertilizers. Phosphate rock, $Ca_3(PO_4)_2$ is available but is too insoluble to be an efficient fertilizer. Therefore it is frequently converted into the more soluble calcium dihydrogen phosphate, generally by the use of sulphuric acid H_2SO_4, the result is hydrated to turn the calcium sulphate into the dihydrate gypsum and sold as "superphosphate of lime"

$$Ca_3(PO_4)_2 + 2H_2SO_4 + 2H_2O \longrightarrow 2CaSO_4 \cdot 2H_2O + Ca(H_2PO_4)_2 \cdot H_2O$$

Alternately phosphate rock may be treated with phosphoric acid to produce a purer form of calcium dihydrogen phosphate and is sold as "triple phosphate".

$$Ca_3(PO_4)_2 + 4\ H_3PO_4 \longrightarrow 3Ca(H_2PO_4)_2$$

Leavening Agent

Monocalcium Phosphate is also used in the food industry as a Leavening agent to cause baked goods to rise. Since it is acidic in nature it provides the better balance towards the other common leavening agent Baking soda (Sodium Bicarbonate) that is alkalic in nature.

This will be useful to adjust the pH of the final end product. Apart from acting as leavening agent, it also possess some microbial inhibition because of the phosphate ions present in the molecule.

Calcium Nitrate

Calcium nitrate, also called Norgessalpeter (Norwegian saltpetre) and Kalksalpeter, is a white coloured soluble salt with the formula $Ca(NO_3)_2$. It is commonly found as a monohydrate, with the formula $Ca(NO_3)_2.H_2O$ and a molecular weight of 182.11 g/mol.

It is deliquescent, readily absorbing moisture from the air. It is used as a reagent, a pyrotechnic and a fertilizer. The fertilizer grade (15.5-0-0 + 19 per cent Ca) is popular in the greenhouse and hydroponics trades; it has some ammonium nitrate and water incorporated in the molecular structure as $5Ca(NO_3)_2.NH_4NO_3.10H_2O$. There also is a formulation without ammonia: $Ca(NO_3)_2.3H_2O$ (12.8-0-0 + 18.3 per cent Ca).

Norgessalpeter was the first nitrogen fertilizer compound to be manufactured. Production began at Notodden, Norway in 1905. Most of the world's calcium nitrate is now made in Porsgrunn.

Nitrocalcite is a form of calcium nitrate which occurs in nature. It occasionally forms an efflorescence where manure contacts concrete or limestone in a dry environment, as in stables or caverns.

Critical Relative Humidity(CRH)

The critical relative humidity (CRH) of a salt is defined as the relative humidity of the surrounding atmosphere (at a certain temperature) at which the material begins to absorb moisture from the atmosphere and below which it will not absorb atmospheric moisture.

When the humidity of the atmosphere is equal to (or is greater than) the critical relative humidity of a sample of salt, the sample will take up water until all of the salt is dissolved to yield a saturated solution.

All water-soluble salts and mixtures have characteristic critical humidities; it is an unique material property.

The critical relative humidity of most salts decreases with increasing temperature. For instance, the critical relative humidity of ammonium nitrate decreases 22 per cent with a temperature from 0°C to 40°C (32°F to 104°F).

The critical relative humidity of several fertilizer salts:

Salt	*Critical Relative Humidity (per cent)*
Calcium nitrate	46.7
Ammonium nitrate	59.4
Sodium nitrate	72.4
Urea	72.5
Ammonium chloride	77.2
Ammonium sulphate	79.2
Diammonium phosphate	82.5
Potassium chloride	84.0
Potassium nitrate	90.5

Contd...

Salt	*Critical Relative Humidity (per cent)*
Monoammonium phosphate	91.6
Monocalcium phosphate	93.6
Potassium sulphate	96.3

Mixtures of salts usually have lower critical humidities than either of the constituents.

Table: Critical relative humidities of pure salts at 30°C (values are per cent relative humidity)

—	*Ammonium nitrate*	*Urea*	*Ammonium sulphate*	*Potassium chloride*
Ammonium nitrate	—	18.1	62.3	67.9
Urea	18.1	—	56.4	60.3
Ammonium sulphate	62.3	56.4	—	71.3
Potassium chloride	67.9	60.3	71.3	—

The effect of salt mixing is most dramatic in the case of ammonium nitrate with urea. This mixture has an extremely low critical relative humidity and can therefore only be used in liquid fertilizers (so called UAN-solutions).

4

Utility of Fermentation

The use of fermentation is an important process in the industry. Though fermentation can have stricter definitions, when speaking of it in Industrial fermentation, it more loosely refers to the breakdown of organic substances and reassembly into other substances. Somewhat paradoxically, fermenter culture in industrial capacity often refers to highly oxygenated and aerobic growth conditions, whereas fermentation in the biochemical context is a strictly anaerobic process.

Food Fermentations

Ancient fermented food processes, such as making bread, wine, cheese, curds, idli, dosa, etc., some of which are some 6,000 yr old, and developed long before man had any knowledge of the existence of the microorganisms involved. Also, fermentation is a powerful weapon for semi-industrialised countries like Honduras, in their willingness to produce biodiesel.

Cheese Making

Pasteurised milk is fermented by the bacteria, *Lactobacillus* sp., which work best at 40 degrees Celsius. *Lactobacillus* works on lactose in milk to coagulate/curdle forming curds. The enzyme rennin is added to milk together with bacteria to cause the milk to coagulate more than it would with just the bacteria.

The curds and fats are then separated from the liquid part of the milk and are pressed to form cheese. After curdling, the liquid part of the milk (whey) is removed and used for making sweets for animal feeds. Different kinds of cheese are made using different types of milk and different mixtures of bacteria (e.g. *Streptococcus* sp.) working at different temperatures.

Flavouring and Variations

Salt can be added Fungal spores (e.g. *Penicillium* can be added to give flavour and blue streaks like in blue Stilton cheese). Large holes in cheese are made by carbon dioxide produced by bacteria.

Pharmaceuticals and Biotech Industry

There are 5 major groups of commercially important Fermentation:

1. *Microbial cells or Biomass as the product:* E.G. Bakers Yeast, Lactic acid bacillus, Bacillus sp.
2. *Microbial Enzymes:* Catalase, Amylase, Protease, Pectinase, Glucose isomerase, Cellulase, Hemicellulase, Lipase, Lactase, Streptokinase, etc.
3. Microbial metabolites :
   a. *Primary metabolites:* Ethanol, Citric acid, Glutamic acid, Lysine, Vitamins, Polysaccharides, etc.
   b. *Secondary metabolites:* All antibiotic fermentation.
4. *Recombinant products:* Insulin, HBV, Interferon, GCSF, Streptokinase.
5. *Biotransformations:* e.g. Phenyl acetyl carbinol, Steroid Biotransformation.

Nutrient Sources for Industrial Fermentation

Medium for Industrial Fermentations: Any Microbe requires Water, Oxygen, Energy source, Carbon source, Nitrogen source and Micronutrients for the growth. Carbon and Energy source + Nitrogen source + O_2 + other requirements $\longrightarrow$ Biomass + Product + by-products + CO_2 + H_2O + heat

Nutrient	*Raw material*
	Carbon
Glucose	Corn sugar, Starch, Cellulose
Sucrose	Sugarcane, Sugar beet molasses
Lactose	Milk whey
Fats	Vegetable oils
Hydrocarbons	Petroleum fractions
	Nitrogen
Protein	Soybean meal, Cornsteep liquor, Distillers' solubles
Ammonia	Pure ammonia or ammonium salts Urea
Nitrate	Nitrate salts
Phosphorus source	Phosphate salts

Trace Elements: Fe, Zn, Cu, Mn, Mo, Co

Antifoaming Agents: Esters, Fatty acids, Silicones, Sulphonates, Polypropylene.

Buffers: Calcium carbonate, Phosphates.

Growth factors: Some microorganisms cannot synthesise the required cell components themselves and need to be supplemented: e.g. Thiamine, Biotin, Calcium pentothenate.

Precursors: Directly incorporated into the desired product: Phenyl ethylamine into Benzyl penicillin, Phenyl acetic acid into Penicillin G.

Inhibitors: To get the specific products: e.g. Sodium barbital for Rifamycin.

Inducers: Majority of the enzymes are inducible and are synthesised in response of inducers: e.g. Starch for Amylases, Maltose for Pollulanase, Pectin for Pectinase.

Chelators: Chelaters are the chemicals used to avoid the precipitation of metal ions. Chelaters like EDTA, Citric acid, Polyphosphates are used in low concentrations.

Use of Fluidised Bed Reactors

The carbon adsorption following biological treatment was particularly effective in reducing both the BOD and COD to low levels. A fluidised bed reactor is a combination of the most common stirred tank packed bed, continuous flow reactors. It is very important to chemical engineering because of its excellent heat and mass transfer characteristics. In a fluidised bed reactor, the substrate is passed upward through the immobilised enzyme bed at a high velocity to lift the particles. However the velocity must not be so high that the enzymes are swept away from the reactor entirely. This causes high mixing, this type of reactors are highly suitable for the exothermic reactions. It is most often applied in immobilised enzyme catalysis.

Filter Beds (Oxidising Beds)

In older plants and plants receiving more variable loads, trickling filter beds are used where the settled sewage liquor is spread onto the surface and of a deep bed made up of coke (carbonised coal), limestone chips or specially fabricated plastic media. Such media must have high surface areas to support the biofilms that form. The liquor is distributed through perforated rotating arms radiating from a central pivot. The distributed liquor trickles through this bed and is collected in drains at the base. These drains also provide a source of air which percolates up through the bed, keeping it aerobic. Biological films of bacteria, protozoa and fungi form on the media's surfaces and eat or otherwise reduce the organic content. This biofilm is grazed by insect larvae and worms which help maintain an optimal thickness. Overloading of beds increases the thickness of the film leading to clogging of the filter media and ponding on the surface.

Biological Aerated Filters

Biological Aerated (or Anoxic) Filter (BAF) combines filtration with biological carbon reduction, nitrification or denitrification. BAF usually includes a reactor filled with a filter media. The media is either in suspension or supported by a gravel layer at the foot of the filter. The dual purpose of this media is to support highly active biomass that is attached to it and to filter suspended solids. Carbon reduction and ammonia conversion occurs in aerobic mode and sometime achieved in a single reactor while nitrate conversion occurs in anoxic mode. BAF is operated either in upflow or downflow configuration depending on design specified by manufacturer.

Membrane Biological Reactors

Membrane biological reactors (MBR) combines activated sludge treatment with a membrane liquid-solid separation process. The membrane component utilises low pressure microfiltration or ultra filtration membranes and eliminates the need for clarification and tertiary filtration. The membranes are typically immersed in the aeration tank (however some applications utilise a separate membrane tank). One of the key benefits of a membrane bioreactor system is that it effectively overcomes the limitations associated with poor settling of sludge in conventional activated sludge (CAS) processes.

The technology permits bioreactor operation with considerably higher mixed liquor suspended solids (MLSS) concentration than CAS systems, which are limited by sludge settling. The process is typically operated at MLSS in the range of 8,000 -12,000 mg/L, while CAS are operated in the range of 2,000 - 3,000 mg/L. The elevated biomass concentration in the membrane bioreactor process allows for very effective removal of both soluble and particulate biodegradable materials at higher loading rates. Thus increased Sludge Retention Times (SRTs)—usually exceeding 15 days—ensure complete nitrification even under extreme cold weather operating conditions.

The cost of building and operating a MBR is usually higher than conventional wastewater treatment, however, as the technology has become increasingly popular and has gained wider acceptance throughout the industry, the life-cycle costs have been steadily decreasing. As well, in developed urban areas where the footprint of the treatment plant is considered a limiting factor MBR Facilities can be considered a desirable option.

Secondary Sedimentation

The final step in the secondary treatment stage is to settle out the biological floc or filter material and produce sewage water containing very low levels of organic material and suspended matter.

Rotating Biological Contactors

Rotating biological contactors (RBCs) are mechanical secondary treatment systems, which are robust and capable of withstanding surges in organic load. RBC's were first installed in Germany in 1960 and have since been developed and refined into a reliable operating unit. The rotating disks support the growth of bacteria and microorganisms present in the sewage, which breakdown and stabilise organic pollutants.

To be successful, microorganisms need both oxygen to live and food to grow. Oxygen is obtained from the atmosphere as the disks rotate. As the microorganisms grow, they build up on the media until they are sloughed off due to shear forces provided by the rotating discs in the sewage. Effluent from the RBC is then passed through final clarifiers where the microorganisms in suspension settle as a sludge. The sludge is withdrawn from the clarifier for further treatment.

Tertiary Treatment

Tertiary treatment provides a final stage to raise the effluent quality before it is discharged to the receiving environment (sea, river, lake, ground, etc.) More than one tertiary treatment process may be used at any treatment plant. If disinfection is practised, it is always the final process. It is also called "effluent polishing".

Filtration

Sand filtration removes much of the residual suspended matter. Filtration over activated carbon removes residual toxins.

Lagooning

Lagooning provides settlement and further biological improvement through storage in large man-made ponds or lagoons. These lagoons are highly aerobic and colonisation by native macrophytes, especially reeds, is often encouraged. Small filter feeding invertebrates such as Daphnia and species of Rotifera greatly assist in treatment by removing fine particulates.

Constructed Wetlands

Constructed wetlands include engineered reedbeds and a range of similar methodologies, all of which provide a high degree of aerobic biological improvement and can often be used instead of secondary treatment for small communities. One example is a small reedbed used to clean the drainage from the elephants' enclosure at Chester Zoo in England.

Waste Removal

Wastewater may contain high levels of the nutrients nitrogen and phosphorus. Excessive release to the environment can lead to a build up of nutrients, called eutrophication, which can in turn encourage the overgrowth of weeds, algae, and cyanobacteria (blue-green algae). This may cause an algal bloom, a rapid growth in the population of algae. The algae numbers are unsustainable and eventually most of them die.

The decomposition of the algae by bacteria uses up so much of oxygen in the water that most or all of the animals die, which creates more organic matter for the bacteria to decompose. In addition to causing deoxygenation, some algal species produce toxins that contaminate drinking water supplies. Different treatment processes are required to remove nitrogen and phosphorus.

Nitrogen Removal

The removal of nitrogen is effected through the biological oxidation of nitrogen from ammonia (nitrification) to nitrate, followed by denitrification, the reduction of nitrate to nitrogen gas. Nitrogen gas is released to the atmosphere and thus removed from the water.

Nitrification itself is a two-step aerobic process, each step facilitated by a different type of bacteria. The oxidation of ammonia (NH_3) to nitrite (NO_2^-) is most often facilitated by *Nitrosomonas* spp. (nitroso=ammonium). Nitrite oxidation to nitrate (NO_3^-), though traditionally believed to be facilitated by *Nitrobacter* spp. (nitro=nitrite), is now known to be facilitated in the environment almost exclusively by *Nitrospira* spp.

Denitrification requires anoxic conditions to encourage the appropriate biological communities to form. It is facilitated by a wide diversity of bacteria. Sand filters, lagooning and reed beds can all be used to reduce nitrogen, but the activated sludge process (if designed well) can do the job the most easily. Since denitrification is the reduction of nitrate to dinitrogen gas, an electron donor is needed. This can be, depending on the wastewater, organic matter (from faeces), sulphide, or an added donor like methanol.

Sometimes the conversion of toxic ammonia to nitrate alone is referred to as tertiary treatment.

Phosphorus Removal

Phosphorus can be removed biologically in a process called enhanced biological phosphorus removal. In this process, specific bacteria, called polyphosphate accumulating organisms, are selectively enriched and accumulate large quantities of phosphorus within their cells (up to 20 per cent of their mass). When the biomass enriched in these bacteria is separated from the treated water, these biosolids have a high fertilizer value.

Phosphorus removal can also be achieved by chemical precipitation, usually with salts of iron (e.g. ferric chloride) or aluminium (e.g. alum). The resulting chemical sludge is difficult to handle and the added chemicals can be expensive. Despite this, chemical phosphorus removal requires significantly smaller equipment footprint than biological removal, is easier to operate and can be more reliable in areas that have wastewater compositions that make biological phosphorus removal difficult.

Disinfection

The purpose of disinfection in the treatment of wastewater is to substantially reduce the number of microorganisms in the water to be discharged back into the environment. The effectiveness of disinfection depends on the quality of the water being treated (e.g., cloudiness, pH, etc.), the type of disinfection being used, the disinfectant dosage (concentration and time), and other environmental variables. Cloudy water will be treated less successfully since solid matter can shield organisms, especially from ultraviolet light or if contact times are low. Generally, short contact times, low doses and high flows all militate against effective disinfection. Common methods of disinfection include ozone, chlorine, or ultraviolet light. Chloramine, which is used for drinking water, is not used in wastewater treatment because of its persistence.

Chlorination remains the most common form of wastewater disinfection in North America due to its low cost and long-term history of effectiveness. One disadvantage is that chlorination of residual organic material can generate chlorinated-organic compounds that may be carcinogenic or harmful to the environment. Residual chlorine or chloramines may also be capable of chlorinating organic material in the natural aquatic environment. Further, because residual chlorine is toxic to aquatic species, the treated effluent must also be chemically dechlorinated, adding to the complexity and cost of treatment.

Ultraviolet (UV) light can be used instead of chlorine, iodine, or other chemicals. Because no chemicals are used, the treated water's taste is more natural and pure as compared to other methods. UV radiation causes damage to the genetic structure of bacteria, viruses, and other pathogens, making them incapable of reproduction. The key disadvantages of UV disinfection are the need for frequent lamp maintenance and replacement and the need for a highly treated effluent to ensure that the target microorganisms are not shielded from the UV radiation (i.e., any solids present in the treated effluent may protect microorganisms from the UV light). In the United Kingdom, light is becoming the most common means of disinfection because of the concerns about the impacts of chlorine in chlorinating residual organics in the wastewater and in chlorinating organics in the receiving water. Edmonton, Alberta, Canada also uses UV light for its water treatment.

Ozone O_3 is generated by passing oxygen O_2 through a high voltage potential resulting in a third oxygen atom becoming attached and forming O_3. Ozone is very unstable and reactive and oxidises most organic material it comes in contact with, thereby destroying many pathogenic microorganisms. Ozone is considered to be safer than chlorine because, unlike chlorine which has to be stored on site (highly poisonous in the event of an accidental release), ozone is generated onsite as needed. Ozonation also produces fewer disinfection by-products than chlorination. A disadvantage of ozone disinfection is the high cost of the ozone generation equipment and the requirements for highly skilled operators.

Package Plants and Batch Reactors

In order to use less space, treat difficult waste, deal with intermittent flow or achieve higher environmental standards, a number of designs of hybrid treatment plants have been produced. Such plants often combine all or at least two stages of the three main treatment stages into one combined stage. In the UK, where a large number of sewage treatment plants serve small populations, package plants are a viable alternative to building discrete structures for each process stage.

The most advanced packaged treatment plant according to a study done by the University of California at Davis for the California State Water Resources Control Board for treating waste and nutrients (phosphorus and nitrogen) in one step economically is the USBF (Upflow Sludge Blanket Filter).

The USBF™ process is a modification of the conventional activated sludge process that incorporates an anoxic selector zone and an upflow sludge blanket filtration clarifier all in one integrated bioreactor vessel. The treatment includes efficient reduction of BOD5 and TSS but also biological nutrient removal (BNR) by the processes of denitrification and "biological luxury uptake". The ensuing compact, modular system takes up less space and contains very few moving parts. The result is an efficient, highly affordable wastewater treatment plant with low maintenance and operating costs.

USBF technology has no inherent capacity limits and is used in a wide range of applications from subdivisions (Bay Point, FL) resorts (Sun Peaks, BC, Kicking Horse, BC) and municipalities (Pahrump, NV), to agricultural and industrial sites. Plants can be retrofitted and expanded from existing sites reducing capital costs. Since there are no mechanical parts and no chemicals needed, operations cost are much less than sequencing batch reactor and membrane bioreactor systems.

Another type of process which combines secondary treatment and settlement is the sequencing batch reactor (SBR). Typically, activated sludge is mixed with raw incoming sewage and mixed and aerated. The resultant mixture is then allowed to settle producing a high quality effluent. The settled sludge is run off and re-aerated before a proportion is returned to the head of the works. SBR plants are now being deployed in many parts of the world including North Liberty, Iowa, and Llanasa, North Wales.

The disadvantage of such processes is that precise control of timing, mixing and aeration is required. This precision is usually achieved by computer controls linked to many sensors in the plant. Such a complex, fragile system is unsuited to places where such controls may be unreliable, or poorly maintained, or where the power supply may be intermittent.

Package plants may be referred to as *high charged* or *low charged*. This refers to the way the biological load is processed. In high charged systems, the biological stage is presented with a high organic load and the combined floc and organic material is then oxygenated for a few hours before being charged again with a new load. In the low charged system the biological stage contains a low organic load and is combined with floculate for a relatively long time.

Sludge Treatment and Disposal

The sludges accumulated in a wastewater treatment process must be treated and disposed of in a safe and effective manner. The purpose of digestion is to reduce the amount of organic matter and the number of disease-causing microorganisms present in the solids. The most common treatment options include anaerobic digestion, aerobic digestion, and composting.

The choice of a wastewater solid treatment method depends on the amount of solids generated and other site-specific conditions. However, in general, composting is most often applied to smaller-scale applications followed by aerobic digestion and then lastly anaerobic digestion for the larger-scale municipal applications.

Anaerobic Digestion

Anaerobic digestion is a bacterial process that is carried out in the absence of oxygen. The process can either be *thermophilic* digestion, in which sludge is fermented in tanks at a temperature of 55°C, or *mesophilic*, at a temperature of around 36°C. Though allowing shorter retention time (and thus smaller tanks), thermophilic digestion is more expensive in terms of energy consumption for heating the sludge.

One major feature of anaerobic digestion is the production of biogas, which can be used in generators for electricity production and/or in boilers for heating purposes.

Aerobic Digestion

Aerobic digestion is a bacterial process occurring in the presence of oxygen. Under aerobic conditions, bacteria rapidly consume organic matter and convert it into carbon dioxide. The operating costs are characteristically much greater than for anaerobic digestion because of the energy costs needed to add oxygen to the process.

Composting

Composting is also an aerobic process that involves mixing the wastewater solids with sources of carbon such as sawdust, straw or wood chips. In the presence of oxygen, bacteria digest both the wastewater solids and the added carbon source and, in doing so, produce a large amount of heat.

Thermal Depolymerisation

Thermal depolymerisation uses hydrous pyrolysis to convert reduced complex organics to oil.

Sludge Disposal

When a liquid sludge is produced, further treatment may be required to make it suitable for final disposal. Typically, sludges are thickened (dewatered) to reduce the volumes transported off-site for disposal. There is no process which completely eliminates the need to dispose of biosolids. There is, however, an additional step some cities are taking to superheat the wastewater sludge and convert it into small pelletised granules that are high in nitrogen and other organic materials. This product is then sold to local farmers and turf farms as a soil amendment or fertilizer, reducing the amount of space required to dispose of sludge in landfills.

Treatment in the Receiving Environment

Many processes in a wastewater treatment plant are designed to mimic the natural treatment processes that occur in the environment, whether that environment is a natural water body or the ground. If not overloaded, bacteria in the environment will consume organic contaminants, although this will reduce the levels of oxygen in the water and may significantly change the overall ecology of the receiving water. Native bacterial populations feed on the organic contaminants, and the numbers of disease-causing microorganisms are reduced by natural environmental conditions such as predation exposure to ultraviolet radiation, for example. Consequently, in cases where the receiving environment provides a high level of dilution, a high degree of wastewater treatment may not be required.

However, recent evidence has demonstrated that very low levels of certain contaminants in wastewater, including hormones (from animal husbandry and residue from human birth control pills) and synthetic materials such as phthalates that mimic hormones in their action, can have an unpredictable adverse impact on the natural biota and potentially on humans if the water is reused for drinking water. In the US and EU, uncontrolled discharges of wastewater to the environment are not permitted under law, and strict water quality requirements are to be met. A significant threat in the coming decades will be the increasing uncontrolled discharges of wastewater within rapidly developing countries.

Sewage Treatment in Developing Countries

There are few reliable figures on the share of the wastewater collected in sewers that is being treated in the world. In many developing countries the bulk of domestic and industrial wastewater is discharged without any treatment or after primary treatment only. In Latin America about 15 per cent of collected wastewater passes through treatment plants (with varying levels of actual treatment). In Venezuela, a below average country in South America with respect to wastewater treatment, 97 per cent of the country's sewage is discharged raw into the environment Even a highly industrialised country such as the People's Republic of China discharges about 55 per cent of all sewage without treatment of any type In a relatively developed Middle Eastern country such as Iran, Tehran's majority of population has totally untreated sewage injected to the city's groundwater. Most of sub-Saharan Africa is without wastewater treatment.

Water utilities in developing countries are chronically underfunded because of low water tariffs, the inexistence of sanitation tariffs in many cases, low billing efficiency (i.e. many users that are billed do not pay) and poor operational efficiency (i.e. there are overly high levels of staff, there are high physical losses, and many users have illegal connections and are thus not being billed). In addition, wastewater treatment typically is the process within the utility that receives the least attention, partly because enforcement of environmental standards is poor. As a result of all these factors, operation and maintenance of many wastewater treatment plants is poor. This is evidenced by the frequent breakdown of equipment, shutdown of electrically operated equipment due to power outages or to reduce costs, and sedimentation due to lack of sludge removal. Developing countries as diverse as Egypt, Algeria, China or Colombia have invested substantial sums in wastewater treatment without achieving a significant impact in terms of environmental improvement. Even if wastewater treatment plants are properly operating, it can be argued that the environmental impact is limited in cases where the assimilative capacity of the receiving waters (ocean with strong currents or large rivers) is high, as it is often the case.

Benefits of Wastewater Treatment Compared to Benefits of Sewage Collection in Developing Countries

Waterborne diseases that are prevalent in developing countries, such as diarrhoea, typhus and cholera, are caused primarily by poor hygiene practices and the absence of improved household sanitation facilities. The public health impact of the discharge of untreated wastewater is comparatively much lower. Hygiene promotion, onsite sanitation and low-cost sanitation thus are likely to have a much greater impact on public health than wastewater treatment. Given the scarcity of financial resources in developing countries and the poor track record of wastewater treatment plants, it could thus be argued that investments should first be undertaken to evacuate wastewater from human settlements and to promote good hygiene practices. Only once this has been achieved, substantial funds should be invested in wastewater treatment plants.

Treatment of Sewage

Sewage treatment, or domestic wastewater treatment, is the process of removing contaminants from wastewater, both run-off and domestic. It includes physical, chemical and biological processes to remove physical, chemical and biological contaminants. Its objective is to produce a waste stream (or treated effluent) and a solid waste or sludge also suitable for discharge or reuse back into the environment. This material is often inadvertently contaminated with toxic organic and inorganic compounds.

Sewage is created by residences, institutions, and commercial and industrial establishments. It can be treated close to where it is created (in septic tanks or on site package plants and other aerobic treatment systems), or collected and transported via a network of pipes and pump stations to a municipal treatment plant.

Sewage collection and treatment is typically subject to local, state and federal regulations and standards (regulation and controls). Industrial sources of wastewater often require specialised treatment processes.

Typically, sewage treatment involves three stages, called *primary, secondary* and *tertiary-treatment*. First, the solids are separated from the wastewater stream. Then dissolved biological matter is progressively converted into a solid mass by using indigenous, water-borne bacteria.

Finally, the biological solids are neutralised then disposed of or reused, and the treated water may be disinfected chemically or physically (for example by lagoons and microfiltration). The final effluent can be discharged into a stream, river, bay, lagoon or wetland, or it can be used for the irrigation of a golf course, green way or park. If it is sufficiently clean, it can also be used for groundwater recharge.

Description

Raw influent (sewage) is the liquid waste from toilets, baths, showers, kitchens, sinks, etc. Household waste that is disposed of via sewers. In many areas sewage also includes some liquid waste from industry and commerce. In the UK, the waste from toilets is termed foul waste, the waste from items such as basins, baths, kitchens is termed sullage water, and the industrial and commercial waste is termed trade waste.

The division of household water drains into greywater and blackwater is becoming more common in the developed world, with greywater being permitted to be used for watering plants or recycled for flushing toilets. A lot of sewage also includes some surface water from roofs or hard-standing areas. Municipal wastewater therefore includes residential, commercial, and industrial liquid waste discharges, and may include stormwater run-off.

Sewage systems capable of handling stormwater are known as combined systems. Such systems are usually avoided since they complicate and thereby reduce the efficiency of sewage treatment plants owing to their seasonality. The variability in flows also leads to often larger than necessary, and subsequently more expensive, treatment facilities.

In addition, heavy storms that contribute more flows than the treatment plant can handle may overwhelm the sewage treatment system, causing a spill or overflow (called a combined sewer overflow, or CSO, in the US). It is preferable to have a separate storm drain system for stormwater in areas that are developed with sewer systems.

The construction of combined sewers is a less common practice in the United States and Canada than in the past and is no longer accepted within building regulations in the UK and other European countries. Instead, liquid waste and stormwater are collected and conveyed in separate sewer systems, referred to as sanitary sewers and storm sewers in the US and as foul sewers and surface water sewers in the UK.

As rainfall runs over the surface of roofs and the ground, it may pick up various contaminants including soil particles, (sediment), heavy metals, organic compounds, animal waste, and oil and grease. Some jurisdictions require stormwater to receive some level of treatment before being discharged directly into waterways. Examples of treatment processes used for stormwater include sedimentation basins, wetlands, buried concrete vaults with various kinds of filters, and vortex separators (to remove coarse solids).

The site where the raw wastewater is processed before it is discharged back to the environment is called a wastewater treatment plant (WWTP). The order and types of mechanical, chemical and biological systems that comprise the wastewater treatment plant are typically the same for most developed countries:

- Mechanical Treatment:
  - Influx (Influent);
  - Removal of large objects;
  - Removal of sand and grit;
  - Preprecipitation.
- Biological Treatment:
  - Oxidation bed (oxidising bed) or aeration system;
  - Post precipitation;
  - Effluent.
- Chemical treatment (this step is usually combined with settling and other processes to remove solids, such as filtration. The combination is referred to in the US as physical-chemical treatment).

Treatment Stages

Primary Treatment: Primary treatment removes the materials that can be easily collected from the raw wastewater and disposed of. The typical materials that are removed during primary treatment include to fats, oils, and greases (also referred to as FOG), sand, gravels and rocks (also referred to as grit), larger settleable solids including human waste, and floatable materials. This step is done entirely with machinery, hence the name mechanical treatment.

Influx (influent) and Removal of Large Objects

In the mechanical treatment, the influx (influent) of sewage water is strained to remove all large objects that are deposited in the sewer system, such as rags, sticks, condoms, sanitary towels (sanitary napkins) or tampons, cans, fruit, etc. This is most commonly done with a manual or automated mechanically raked screen. This type of waste is removed because it can damage or clog the equipment in the sewage treatment plant.

Sand and Grit Removal

Primary treatment typically includes a sand or grit channel or chamber where the velocity of the incoming wastewater is carefully controlled to allow sand grit and stones to settle, while keeping the majority of the suspended organic material in the water column. This equipment is called a detritor or sand catcher. Sand grit and stones need to be removed early in the process to avoid damage to pumps and other equipment in the remaining treatment stages. Sometimes there is a sand washer (grit classifier) followed by a conveyor that transports the sand to a container for disposal. The contents from the sand catcher may be fed into the incinerator in a sludge processing plant, but in many cases, the sand and grit is sent to a land-fill.

Screening and Maceration (Raw Sewage Pumping)

The grit free liquid is then passed through fixed or rotating screens to remove floating and larger material like rags and smaller particulates such as undigested food particles or materials that were not adequately pulverised with residential food disposal units. Screenings are collected and may be returned to the sludge treatment plant or may be disposed of off-site by landfilling or incineration. Maceration, in which solids are cut into small particles through the use of rotating knife edges mounted on a revolving cylinder, is often used in plants that do not remove the larger particulates at this stage of treatment and that can process this particulate waste.

Sedimentation

Many plants have a sedimentation stage where the sewage is allowed to pass slowly through large tanks, commonly called "primary clarifiers" or "primary sedimentation tanks". The tanks are large enough that faecal solids can settle and floating material such as grease and oils can rise to the surface and be skimmed off.

The main purpose of the primary stage is to produce a generally homogeneous liquid capable of being treated biologically and a sludge that can be separately treated or processed. Primary settlement tanks are usually equipped with mechanically driven scrapers that continually drive the collected sludge towards a hopper in the base of the tank from where it can be pumped to further sludge treatment stages.

Secondary Treatment

Secondary treatment is designed to substantially degrade the biological content of the sewage such as are derived from human waste, food waste, soaps and detergent. The majority of municipal and industrial plants treat the settled sewage liquor using aerobic biological processes.

For this to be effective, the biota require both oxygen and a substrate on which to live. There are number of ways in which this is done. In all these methods, the bacteria and protozoa consume biodegradable soluble organic contaminants (e.g. sugars, fats, organic short-chain carbon molecules, etc.) and bind much of the less soluble fractions into floc. Secondary treatment systems are classified as fixed film or suspended growth. Fixed-film treatment process including trickling filters and rotating biological contactors where the biomass grows on media and the sewage passes over its surface. In suspended growth systems—such as activated sludge—the biomass is well mixed with the sewage and can be operated in a smaller space than fixed-film systems that treat the same amount of water. However, fixed-film systems are more able to cope with drastic changes in the amount of biological material and can provide higher removal rates for organic material and suspended solids than suspended growth systems.

Roughing filters are intended to treat particularly strong or variable organic loads, typically industrial, to allow them to be treated by conventional secondary treatment processes. Characteristics include typically tall, circular filters filled with open synthetic filter media to which wastewater is applied at a relatively high rate. Designed to allow high hydraulic loading and a high flow-through of air. On larger installations, air is forced through the media using blowers. The resultant wastewater is usually within the normal range for conventional treatment processes.

Activated Sludge

Activated sludge plants use a variety of mechanisms and processes to use dissolved oxygen to promote the growth of biological floc that substantially removes organic material. It also traps particular material and can, under ideal conditions, convert ammonia to nitrite and nitrate ultimately to nitrogen gas.

5

Chemistry's Application in Bio-organics

Components of Nucleic Acids

Nucleic acids: Nucleic acids are bio molecules which are found in the nuclei of all living cells in form of nucleoproteins.

They are biopolymers in which the repeating structural unit or monomeric unit is a nucleotide (polynucleotides). Each nucleotide consists of three components—

1. A pentose sugar, *i.e.*, D(–) ribose or 2-deoxyribose (furanose form).
2. A heterocyclic nitrogenous base, two types—
   (*a*) *purines:* Adenine (A) and Guanine (G).
   (*b*) *pyrimidines*: Uracil (U), Thymine (T), Cytosine (C).
3. *Phosphoric acid*: Nucleotides are the phosphoric acid esters of nucleosides (nucleosides contain only pentose sugar and a nitrogen base).

Two types of nucleic acids are:

Deoxyribonucleic Acid (DNA): Sugar 2-deoxyribose, a phosphate group and a heterocyclic base. The base is either purine A or G and pyrimidine C or T.

Ribonucleic Acid (RNA): Sugar is ribose, a phosphate group and a heterocyclic base. The base is either purine A or G and pyrimidine C or U.

Thus the RNA differ from DNA in sugar (ribose) and the pyrimidine base U replaces T.

The primary structure of nucleic acids refers to the sequence in which the four nitrogen bases (A, G, C and T in DNA and A, G, C and U) in RNA are attached to sugar phosphate backbone of the nucleotide chain.

Secondary structure of DNA consists of two strands of polynucleotides coiled around each there in the form of double helix. The backbone of each strand is sugar-phosphate unit and the base unit of each strand are pointed into the interior of the helix and are linked through H-bonds. G and C are held by three H-bonds, A and T are held by two bonds. Unlike DNA, RNA has a single strand.

Change in Base Sequence in DNA Causing Mutation: A change in single base pair in DNA can cause substitution of a different amino acid in the protein being synthesised (as is seen in sickle cell haemoglobin). In a silent mutation, the substitution does not significantly change the normal activity of protein. A lethal mutation is characterised by production of a defective harmful protein or failure to produce an essential protein.

The H-bonds between the paired bases and the Van der Waals attractive forces that hold the staked base-pairs, aligned one pair on top of the other, are disrupted. The ordered conformation breaks and the double helix dissociates into random, disordered coils. The viscosity of the solution drops sharply.

The double helix structure of DNA has the hydrophobic bases pointing to the centre of the helix in an almost planar arrangement. These base-pairs are closely stacked perpendicular to the long axis of the chain, and are attracted to each other by Van der Waals forces. The hydrophilic phosphates are negatively charged at the pH of the cell and point to the outside.

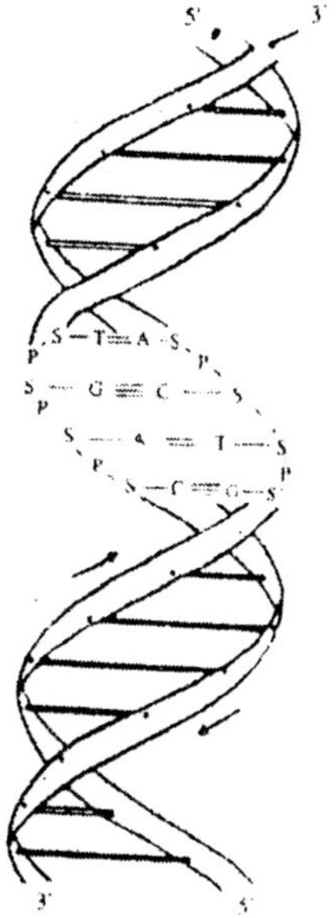

Double-helix structure of DNA.

The two sugar-phosphate chains are aligned in opposite directions. The 5′OH linkage is at the top of the one strand and 3′OH linkage at the top of the other strand. In the figure the number of dashes between the base symbols indicate the number of H-bonds. S is deoxyribose and P is O=P—O$^-$.

DNA directing Protein Synthesis in a Cell: The DNA directed protein synthesis occurs in two steps—

Transcription: It involves copying of DNA base sequences into a *m*RNA. A small portion of DNA double helix unwinds and are of the two DNA strands act as the template for the synthesis of *m*RNA. Ribonucleotides assemble along the uncoiled template in accordance with the base pairing principle. For example, U in the RNA being formed appears opposite A of DNA; C opposite G; A opposite T and G opposite C of DNA. Then occurs the bond formation between the various nucleotides thus assembled.

When *m*RNA strand is synthesised, the DNA-RNA double helix separates. The *m*RNA migrates to the cell cytoplasm while DNA returns to its normal double helix structure. Similarly *t*RNA and *r*RNA are synthesised.

Translation: The *m*RNA directs protein synthesis in the cytoplasm of the cell with the help of *r*RNA and *t*RNA. This process is called translation. The *m*RNA synthesised above gets attached to very small ribosome particles (60% *r*RNA and 40% protein). On the ribosome, *m*RNA serves as the template for protein synthesis.

The specific ribonucleotide sequence in the *m*RNA forms a code that determines the order in which the different amino acid residues have to be joined. The four bases, *i.e.*, A, C, G and U have been shown to act in the form of *triplets*. Since a particular amino acid codes a specific amino acid, therefore, these triplets are called *codons*. Since there are four bases, therefore, $4^3 = 64$ triplets or codons are possible. Out of these 61 for specific amino acids while the remaining 3 (UUA, UGA and UAG) are known to code for chain termination (nonsense or stop codons). Since there are only 20 different amino acids, therefore, more than one triplet combinations code for the same amino acid, *e.g.*, GGU, GGA, GGC, GGG code for glycine and two triplets UUU and UUC code for phenylanine.

*m*RNA recognise the amino acid with the help of *t*RNA so there are 61 different *t*RNA. All the *t*RNA have a CCA triplet which is attached to amino acid by an ester linkage.

At one end of the loop of *t*RNA there is a ribonucleotide triplet called *anticodon* which is complementary to a codon on *m*RNA. Each codon of *m*RNA is read in a serial order by an anticodon of *t*RNA and matched. If matching occurs, the *t*RNA transfers the desired amino acid to the growing polypeptide chain on the ribosome. When the synthesis of a specific protein is completed, a stop codon signals the end and the synthesised protein is released from the ribosome.

On hydrolysis of DNA, it is found that although the ratio of bases vary from one DNA to another, the ratios of C : G and A : T are always 1 : 1.

The C and A is one strand always matches the G and T respectively in other strand. This matching is called base-pairing. (*b*) Base pairing occurs by H-bonding with maximum efficiency between a pyrimidine and a purine base, specifically between A and T and between C and G.

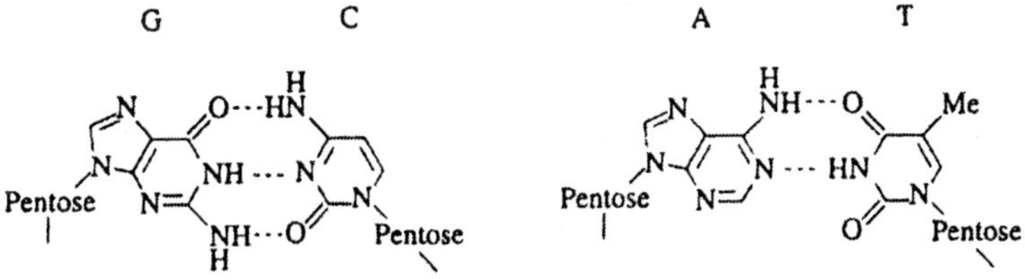

G-C base-pairing A-T base-pairing

This suggest that RNA helixes are single stranded not double helixes like DNA.

An error of a single base in each strand of DNA can bring about the amino acid residue error that causes sickle-cell anemia.

A change from C—T—T to C—A—T or a change from C—T—C to C—A—C.

Tautomeric structures of adenine, guanine, cytosine, thymine and uracil are—

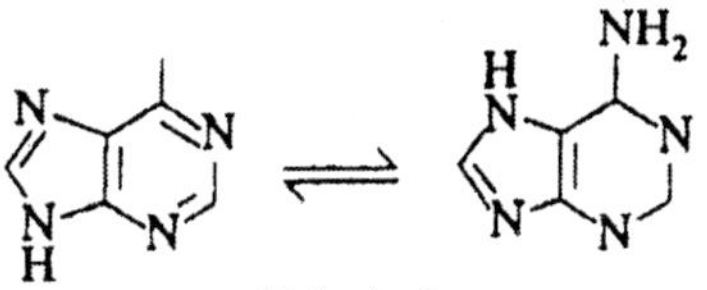

(Adenine)

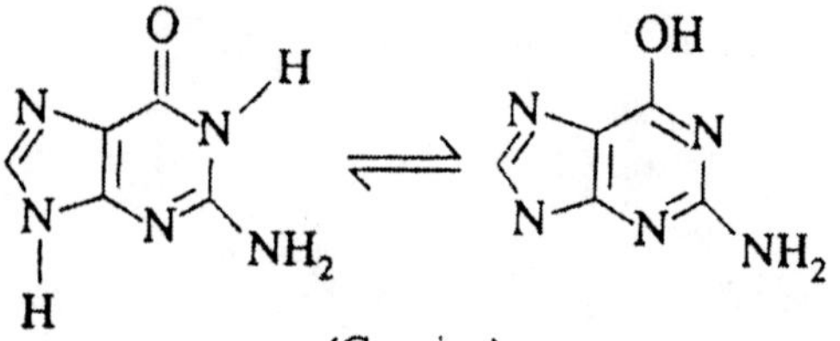

(Guanine)

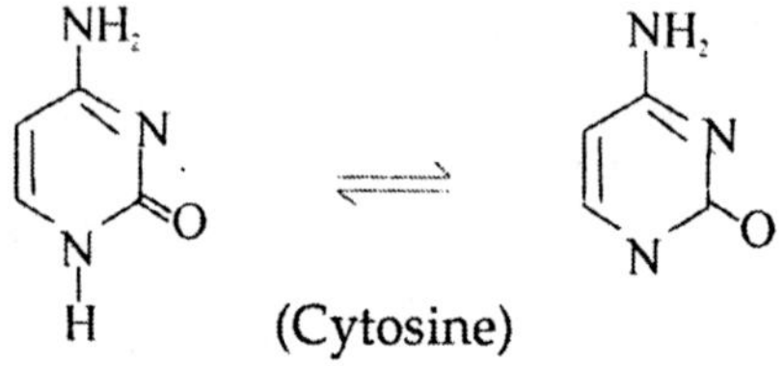

(Cytosine)

(Thymine)

(Uracil)

The successive hydrolytic degradation products of a nucleoprotein outlined.

Nucleoprotein $\longrightarrow$ nucleic acid (+ proteins) $\longrightarrow$ nucleotides $\longrightarrow$ nucleosides (+ phosphate ion) $\longrightarrow$ pentose + heterocyclic base.

Base-pairs in the gene, needed to code for the enzyme lysozyme (129 amino acids) found in egg white.

Three base-pairs code for one amino acid, and two more triplets are required to start and stop signals, or $3 \times 129 + 3 + 2 = 393$ base pairs.

HNO_2 converts —NH_2 to —OH, *i.e.*, cytosine to uracil and guanine to xanthine via diazotisation, thereby changing the genetic code and leading possible mutation.

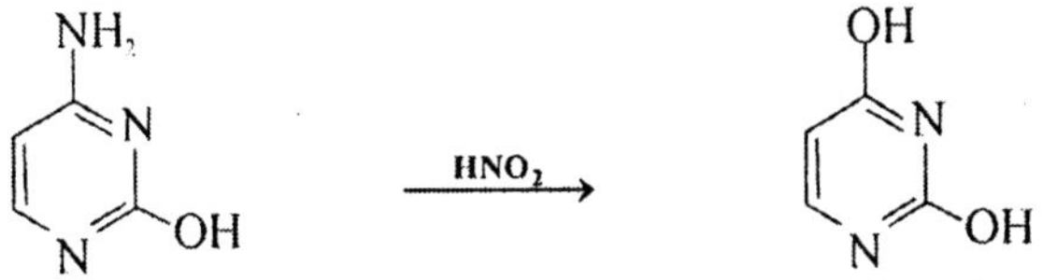

Cytosine Uracil

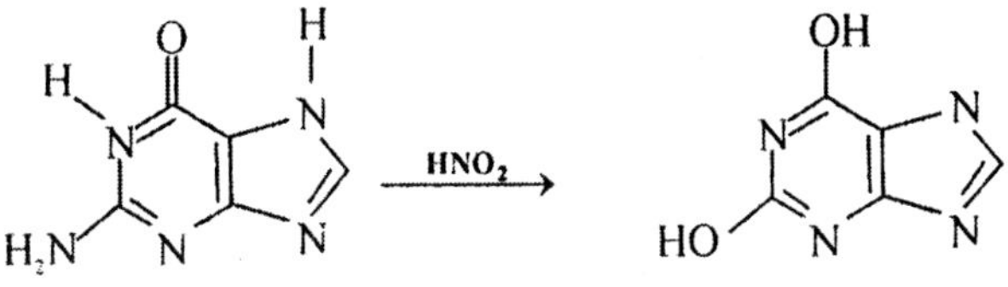

Guanine Xanthine

Replication: The process by which a single DNA molecule produces two identical copies of itself is called replication (cell division). Replication is a enzyme catalysed process. It begins with partial unwinding of the double helix through breaking of H-bonds between pairs of bases leading to the formation of two isolated strands. Each strand then act the template for the synthesis of two new strands. Because of the base pairing principle, each new strand is complementary to its old template strand. As a result, two identical copies of DNA from the parent DNA are produced. Each copy is then passed on the two new cells resulting from cell division. In this way hereditary effects are transmitted from one cell to the other.

ATP Regarded as Store House of Energy: Adenosine triphosphate (ATP) in a nucleotide consists of purine base adenine, a pentose sugar ribose and three molecules of phosphate. It contains two oxygen to phosphorus bonds between two phosphate units. These phosphorus bonds are called high energy phosphatic bonds.

ATP is an energy rich molecule. This is because of the presence of four negatively charged oxygen atoms very close to each other. These four negatively charged O-atoms experience very high repulsive energy. When ATP is hydrolysed, the hydrolysis of bond results in decrease in repulsive forces and consequently, a large amount of energy is released. They hydrolysis of ATP proceeds as—

$$\text{ATP} \xrightarrow{\text{Hydrolysis}} \underset{\text{Adenosine diphosphate}}{\text{ADP}} + \text{Pi}$$

$$\text{ADP} \xrightarrow{\text{Hydrolysis}} \text{AMP} + \text{Pi}$$

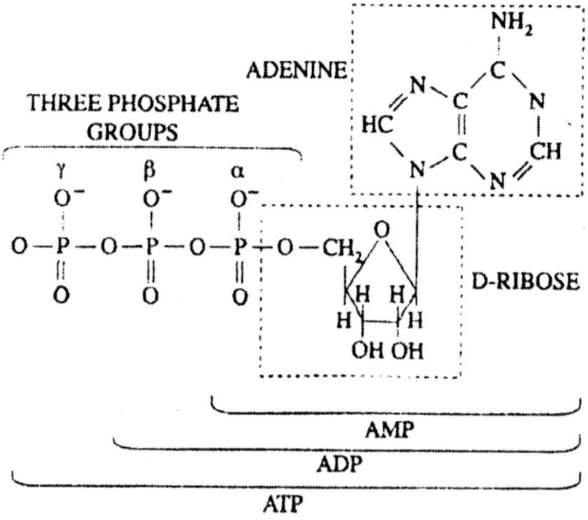

The structure of ATP, ADP and AMP.

The energy producing reactions are coupled to carry out some other chemical reactions which are otherwise not energetically possible. ATP, ADP and AMP occur not only in cell cytosol but also in mitochondria and the nucleus.

During phosphate transfer, ATP breaks down to give AMP or ADP and not adenosine and PPPi. The structure of a mononucleotide.

ATP is an energy rich molecule because its triphosphate unit contains two phosphoanhydride bonds. A large amount of free energy is liberated when ATP is hydrolysed to ADP or AMP.

ATP + $H_2O \rightleftharpoons$ ADP + Pi + H^+ $\Delta G° = -7.3$ k cal/mol

ATP + $H_2O \rightleftharpoons$ AMp + PPi + H^+ $\Delta G° = -7.3$ k cal/mol

ATP is not breakdown to adenosine and PPPi because ATP, AMp and ADP are interconvertible by the enzyme myokinase.

$$\text{ATP + AMP} \rightleftharpoons \text{ADP + ADP}$$

Moreover $\Delta G°$ value of hydrolysis of yield adenosine and phosphate is much lower (about – 3.4 k cal/mol). Thus the phosphate group of AMP (*i.e.*, the α-phosphate group of ATP) is in the lower energy class.

Nucleotides are the phosphoric acid ester of nucleoside, while nucleosides are compounds in which nitrogenous bases (purines and pyrimidines) are conjugated to the pentose sugar (ribose or deoxyribose) by a β-glycosidic linkage. AMP, ATP, ADP, GMP, CMP, UMP are the examples of mononucleotides.

Glyceraldehyde phosphate and dihydroxyacetone phosphate react in the presence of aldolase to yield fructose diphosphate as the only product under physiological conditions.

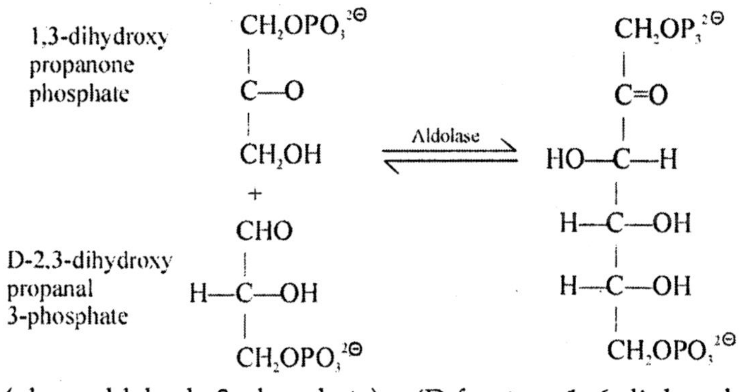

(glyceraldehyde 3-phosphate) (D-fructose-1, 6-diphosphate)

By the reaction of D-glyceraldehyde and 1,3-dihydroxypropane (both as monophosphate ester), D-fructose as the 1,6-diphosphate ester is formed. The process is readily reversible and is catalysed by an enzyme known as aldolase.

This is a type of aldol addition (known as biological aldol addition) and is one of the reaction in the metabolism of carbohydrates by the glycolic pathway.

It seems unlikely that this reaction could occur in quite the same way as in the laboratory aldol reaction, because the enolate anion of the donor molecule (dihydroxypropanone) is not expected to be formed in significant amount of the pH of living cell. In fact there is strong evidence that the enzyme behaves as amino (ENH_2) compound and reacts with carbonyl group of dihydroxy propanone to form an imine. This implies that the imine form of dihydroxy propanone is a key intermediate in the overall aldol-type addition.

The Fundamentals

Primary, Secondary, Tertiary and Quaternary Structure of a Protein.

The Bonding, Greatly Responsible for the Secondary structure.

Primary structure of a protein is simply amino acids sequence of the peptide chain. The secondary structure is a result of the different conformations that the chain can take. The tertiary structure refers to the three dimensional shape that results from twisting, bending and folding of protein helix. The quaternary structure refers to the way in which these amino acid chains of a complex protein are associated with each other (oligomer, dimers, trimers, etc.).

(*a*) The H-bonding between N—H of one amino acid residue and O=C of another properly situated amino acid residue.

(*b*) The peptide sequence is coiled into a right handed spiral in the α-helix, with the R groups positioned on the outside of the spiral. Each amide H—N bonds to the O=C on the next turn of the coil, four residues away by H-bons, stabilising this arrangement.

In the pleated sheet of β-structure, the peptide chains lie side by side in an open structure, with inter-chain amide H-bonding holding the chains together. Parallel pleated sheets have chains running in the same direction, all with their N-terminal residues, starting at the same end. Antiparallel pleated sheets have their chains running in opposite directions. The α C's rotate slightly out of plane of the sheet to minimise repulsions between their bulky R groups, giving rise to the pleats. In both cases, the R groups alternate positions above and below the sheets.

The random coil structure has no repeating geometric pattern; encompassed within it are sequences in a helical conformation, a pleated conformation, and regions that appear to have no discernible repeating structure, but are actually not random conformation.

Bonding Responsible for the Structure of Proteins: The following are the attractive forces responsible for the tertiary structure—

1. Ionic: bonding between COO^- and NH_3^+ at different sites;
2. H-bonding: mainly between side chain nH_2 and COOH, also involving OH's (of serine, for example) and N—H of tryptophane;
3. Weakly hydrophobic Vander Waals attractive forces engendered by side chain R groups;
4. Disulphide cross linkages between loops of the polypeptide chain.

The same kind of attractive and repulsive forces responsible for the tertiary structure operate to hold together and stabilise the subunit of the quaternary structure.

Amino Acids given in the List B with that of Type given in List A. Matched:

	List A		List B
(1)	*Mercapto*	(i)	*Valine*
(2)	*Hydroxy*	(ii)	*Proline*
(3)	*Carboxamide*	(iii)	*Methionine*
(4)	*Unsubstituted*	(iv)	*Tyrosine*
(5)	*Aromatic*	(v)	*Serine*
(6)	*Heterocyclic*	(vi)	*Cysteine*
(7)	*Thioether*	(vii)	*Glutamine*

(1)—(*vi*), (2)—(*v*), (3)—(*vii*), (4)—(*i*), (5)—(*iv*), (6)—(*ii*), (7)—(*iii*).

An elemental analysis of cytochrome c, an enzyme involved in oxidation reduction process, gave 0.43 per cent Fe and 1.48 per cent S. The minimum molecular weight of enzyme and The Minimum Number of Iron Atoms per Molecule of Sulphur Atom.

If one Fe atom per molecule

$$0.43\%\ \text{Fe} = \frac{55.8}{\text{m.w.}} \times 100$$

$$\text{m.w.} = 13{,}000$$

If one S atom per molecule

$$1.48\%\ \text{S} = \frac{32.1}{\text{m.w.}} \times 100$$

$$\text{m.w.} = 2170$$

The minimum m.w. 13,000; 1 Fe and 6 S.

6

Chemistry's Application in Bio-inorganics

Active Site Structure of Rubredoxin

There are several non-haeme iron-sulphur proteins that are involved in electron transfer. They contain distinct iron-sulphur clusters composed of iron atoms, sulphydryl groups from cysteine residues and inorganic or labile sulphur atoms or sulphide ions. The labile sulphur is readily removed by washing with acid. The cysteine moieties are incorporated within the protein chain and are thus not labile. The simplest type of cluster is bacteria *rubredoxin*, (Cys-S)4 Fe (often abbreviated Fe_1S_0 where S stands for inorganic sulphur), 'and contains only non-labile sulphur.

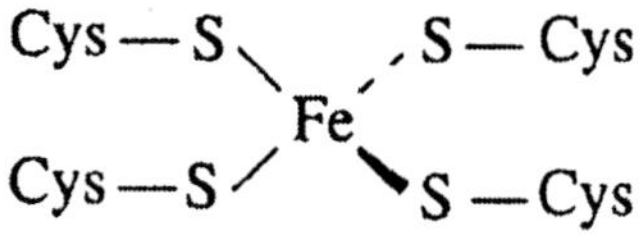

Fig. 1 : Fe_1S_0 in bacterial rubredoxin

It is a bacterial protein of uncertain function with a molecular weight of 6000. The single iron atom is at the centre of a tetrahedron of four cysteine ligands.

Ferredoxin: Two common forms of ferredoxin are photosynthetic ferredoxin and cubane-like ferredoxin.

The cluster in the ferredoxin molecule associated with photosynthesis in higher plants is thought to have the bridged structure Fe_2S_2 as shown in figure. It is known as photosynthetic ferredoxin.

Fig. 2 : Fe_2S_2 in photosynthetic ferredoxin

The cluster found in certain bacterial ferredoxins involved in anaerobic metabolism. It consists of a cubane-like cluster of four iron atoms, four labile sulphur atoms, thus Fe_4S_4, and four cysteine ligands.

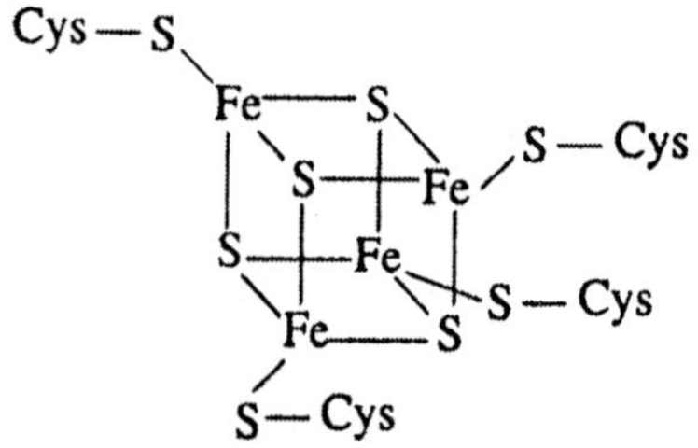

Fe4S4 in cubane like ferredoxin

The Structure of the Proposed Model for the Active Site of Nitrogenase:

Nitrogenase: The best known nitrogen fixing bacterium is Rhizobium. This contains the metalloenzyme nitrogenase. The active enzyme in nitrogen fixation is *nitrogenase.* Nitrogenase contains two proteins, molybdoferredoxin and azoferredoxin. Molybdoferredoxin is brown, air sensitive, contains two Mo atom, 24–36 Fe atoms and 24–36 S atoms together with a protein d_2B_2 tetramer with a and is a molecular weight of about 225000. Azoferredoxin is yellow, air sensitive, is a derivative of ferredoxin $Fe_4S_4(SR)_4$ and has a molecular weight in the range 50000-70000.

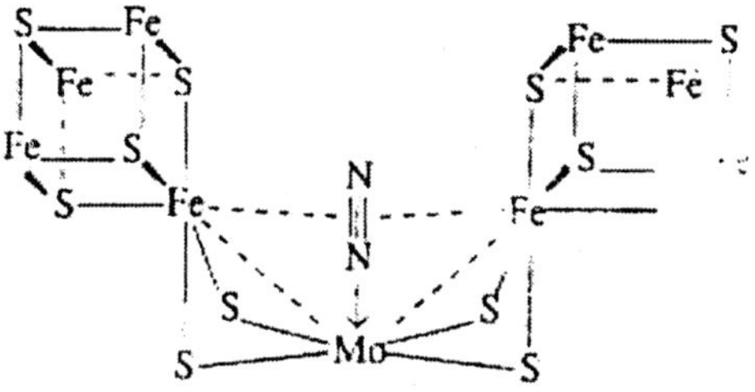

Fig. 4 : A proposed model of the active sites in nitrogenase. The key units are the Mo atom the iron-sulphur structures.

It seems likely that the active site for dinitrogen binding involves the molybdenum atom. It has been established by extended X-ray Absorption Fine Structure (EXFAS) measurements that the coordination sphere consists of several sulphur atoms at distance of about 235 pm. An Mo=O double bond, so common in complexes of Mo (IV) and Mo (VI) is not present. There are other heavy atoms, perhaps iron, nearby (~270 pm).

The iron-sulphur clusters probably act as redox centres. It is not certain exactly how nitrogen fixation occurs. It is thought that N_2 bonds to the Mo in molybdoferredoxin. If the Mo is sufficiently reduced it can probably bond to the antibonding orbitals on N_2. Then the Fe in azoferredoxin is reduced by free ferredoxin. An electron is transferred from reduced azoferredoxin to molybdoferredoxin, possibly via the Fe, then from the Mo to the. N_2-. Protons are then added to N_2 giving NH_3. ATP is required to provide the necessary energy. Nitrogenase also reduces alkynes to alkenes, and cyanide.

Four Steps Involved in Oxidation of Water: The photosynthetic process in green plants consists of splitting the elements of water, followed by reduction of carbon dioxide.

$$2H_20 \longrightarrow 4[H] + O_2$$

$$CO_2 + \frac{x}{2} 4[H] \longrightarrow (CH_2O)_x + \frac{x}{2} O_2$$

where 4[H] does not imply free atoms of hydrogen but a reducing capacity of four equivalents. In all di-oxygen producing organisms, there are two coupled photosynthetic systems PS-I and PS-II. The two differ in the type of chlorophyll present and in the accessory chemicals for processing the trapped energy of photon. The primary product of PS-I is reduced carbon, and the primary product of PS-II is energy in one form of two moles of ATP with molecular oxygen as a chemical by product.

The oxidised form of PS-II is a strong enough oxidising agent to oxidise water. However, the reactions involves the net transfer of four electrons and is achieved only by a complex

$$2H_2O_{(l)} \longrightarrow 4H^+{}_{(aq)} + O_{2\ (g)} + 4e^-$$

series of redox reactions of a manganese based enzyme which transfers electrons to the photochemically active centre. Experiments with successive light flashes show that the enzyme system is oxidised in four one-electron steps before accomplishing the four-electron oxidation of water to O_2. If the redox reservoir involves the reduction of Mn (IV) to Mn (II), the enzyme must contain two Mn atoms to accommodate four electrons. The four steps of oxidation which precede the release of O_2 in the release of O_2 in oxygen evolving system of PS-II is shown in fig.

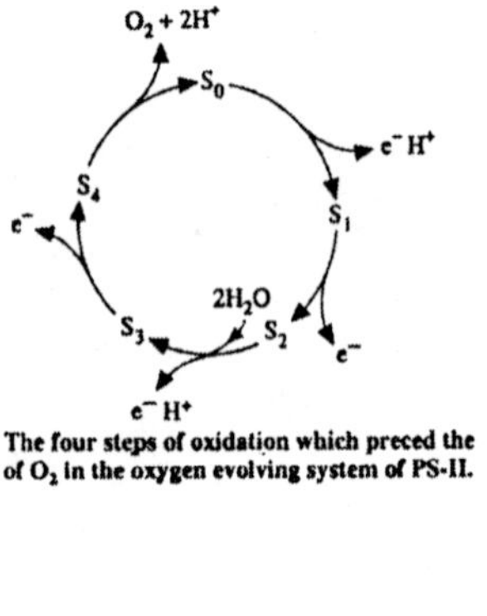

Fig. 5 : The four steps of oxidation which preced the release of O_2 in the oxygen evolving system of PS-II.

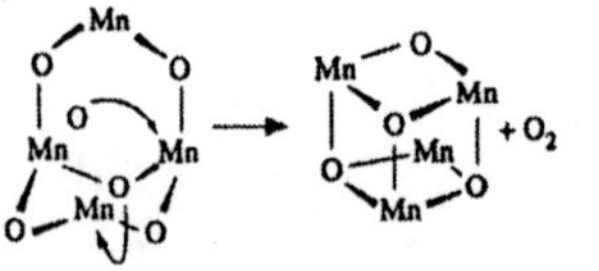

Fig. 6 : S_4 'Adamantane' form S_0 'Cubane' form. A possible analog of the step S_4 to S_0 in previous fig. in which an 'adamantane' form of the enzyme converts to a 'cubane' form with loss of O_2.

Oxygen Uptake and Equilibrium in Myoglobin and Haemoglobin: The function of both haemoglobin and myoglobin is to bind oxygen but their physiological roles are very different. Haemoglobin picks up oxygen in the lungs and carries it to the tissues via the circulatory system.

Cellular oxygen is bound by myoglobin molecules that store it until it is required for metabolic action, where upon they release it to other acceptors. Haemoglobin has a additional function, however, and that is to carry CO_2 back to the lungs; this is done by certain amino acid side chains, and the haeme groups are not directly involved. Because the circumstances under which Hb and Mb are required to bind and release O_2 are very different, the two substances have quite different binding constants as a function of O_2 partial pressure.

Haemoglobin is not simply a passive container of oxygen but an intricate molecular machine. This may be appreciated by comparing its affinity for O_2 to that of myoglobin. For myoglobin (Mb) we have the following simple equilibrium

$$Mb + O_2 \longrightarrow MbO_2$$

$$K = \frac{[MbO_2]}{[Mb][O_2]}$$

If f represents the fraction of myoglobin molecules bearing oxygen and P represents the equilibrium partial pressure of oxygen then

$$K = \frac{f}{(1-f)P} \quad \text{and} \quad f = \frac{KP}{1+KP}$$

This is the equation for the hyperbolic curve shown in Fig. Haemoglobin with its four subunits has more complex behaviour; it approximately follows the equation

$$f = \frac{KP^n}{1+KP^n} \quad n \approx 2{\cdot}8$$

where the exact value of n depends on pH.

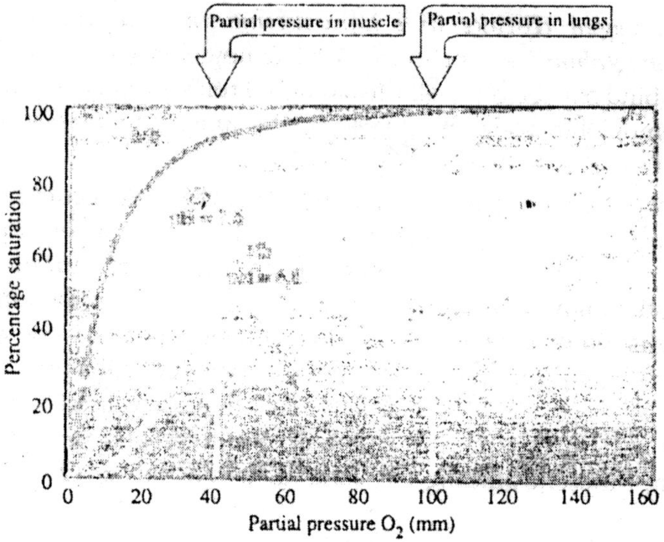

The oxygen-binding curves for myoglobin (Mb) and hemoglobin (Hb), showing also the pH-dependence (Bohr effect) for the latter.

Thus, for haemoglobin (Hb) the oxygen binding curves are sigmoidal as is shown in the above figure. The fact that n exceeds the unit can be ascribed physically to the fact that attachment of O_2 to one haeme group increases the binding constant for the next O_2, which in turn increases the constant for the next one and so on. This is called cooperativity. Just as a value of $n = 1$ would represent the lower limit of no cooperativity a value of $n = 4$ would represent the maximum cooperativity, since it would imply that only Hb and $Hb(O_2)_4$ would be participants in the equilibrium.

Although Hb is about as good an O_2 binder as Mb at high O_2 pressure, it is much porer at the lower pressure prevailing in muscle and, hence, passes on its oxygen to the Mb as required. Moreover, the need for O_2 will be greatest in tissues that have already consumed oxygen and simultaneously have produced CO_2. The CO_2 lowers the pH, thus causing the Hb to release even more O_2 to the Mb. The pH-sensitivity (called the Bohr effect), as well as the progressive increase of the O_2 binding constants in Hb, is due to interactions between the subunits;

Mb behaves more simply because it consists of only one unit. It is clear that each of the two is essential in the complete oxygen transport process.

Important Structural Features of the Active Sites of Metal Centres

Deoxyhaemoglobin, Cytochrome c, Vitamin B_{12}.

Deoxyhaemoglobin: The oxygenated form of haemoglobin is called oxyhaemoglobin and the reduced form is called deoxyhaemoglobin. In oxyhaemoglobin the Fe^{2+} is in the low spin state and is diamagnetic. It is just the right size to fit in the hole at the centre of the porphyrin ring. The porphyrin is both planar and rigid. In deoxyhaemoglobin the Fe^{2+} is in the high spin state and is paramagnetic. The size of Fe^{2+} increases by 28 per cent when it changes from low spin to high spin, *i.e.*, from 0·61 Å to 0·78 Å. Thus in deoxyhaemoglobin the Fe^{2+} is too large to fit in the hole at the centre of the porphyrin, and is situated 0·7-0·8 Å above the ring, thus distorting the bands round the Fe. Thus the presence of O_2 changes the electronic arrangement of Fe^{2+} and also distorts the shape of the complex. The iron atom in deoxyhaemoglobin has square pyramidal coordination.

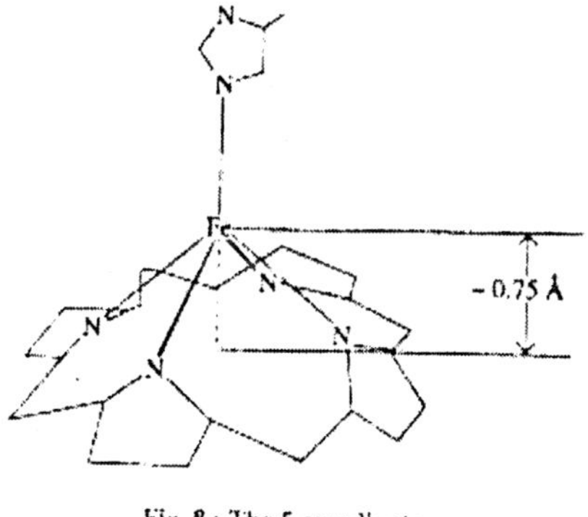

Fig. 8 : The 5-coordinate, high-spin Fe(II) atom in deoxyhemoglobin.

Cytochrome c: There are many cytochromes, which differ in slight detail, but these are broadly grouped together as cytochrome *a*, cytochrome *b* and cytochrome *c*. The active centre of the cytochrome is the haeme group. It consists of a porphyrin ring chelated to an iron atom. The porphyrin ring consists of a macrocyclic pyrrole system with conjugated double bonds and various groups attached to the perimeter.

Cytochrome *c* is an important biological intermediate in electron transfer. This metalloprotein, found in all cells, has a molecular weight of approximately 12,800 and contains 104 amino acids (in vertebrates). It is an electron carrier for oxidative phosphorylation, transferring electrons to O_2. The energy released in this process is used to synthesise ATP. The haeme group of cytochrome *c* lies near the surface of the protein. The iron atom is six coordinate, with bonds to five nitrogen atoms (four from porphyrin, one from a histidine nitrogen) and one sulphur atom from a cysteine. Since all six coordination sites are occupied, direct electron transfer to iron is not possible, and the electron must pass through the surrounding protein bridge work. The importance of cytochrome *c* in photosynthesis and respiration indicates that it is probably one of the oldest of the chemicals involved in biological processes.

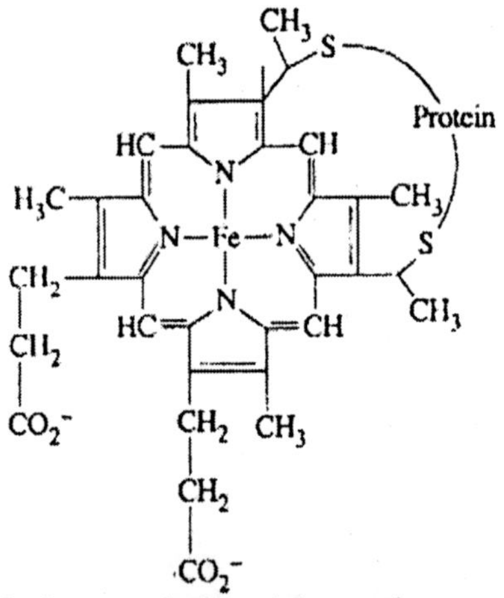

The iron porphyrin unit in cytochrome *c*.

Vitamin B12: Vitamin B12 is an important cobalt complex. The vitamin was isolated from liver after it was found that eating large quantities of raw liver was an effective treatment for pernicious anaemia. The term vitamin B12, refers to cyanocobalamin. Vitamin B12 is a coenzyme, and serves as a prosthetic group which is tightly bound to several enzymes in the body. It is called a biological Grignard reagent. Vitamin B12 is unique in that it appears to be synthesised only by microorganisms especially anaerobic bacteria.

The structure of vitamin B12 complex consists of four principal components—

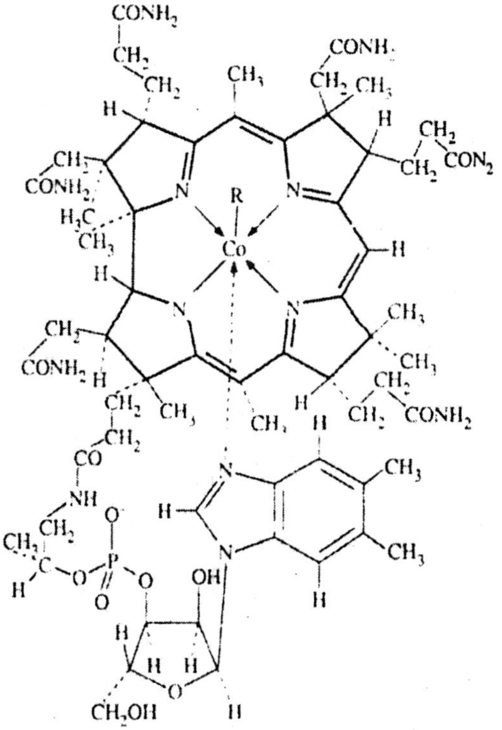

Vitamin B_{12}.
The corrin ring is shown in heavy type.

1. A cobalt (III) ion.
2. A macrocyclic ligand called the *corrin ring,* which bears various substituents. It resembles the porphine ring, but differs in various ways, notably in the absence of one methine, =CH—, bridge between a pair of pyrrole rings.

3. A complex organic portion consisting of a phosphate group, a sugar, and an organic base, the latter being coordinated to the cobalt atom.

4. CN^- ligand coordinated to the cobalt ion.

The Co atom is bonded to four ring N atoms. The fifth position is occupied by another N from a side chain (α-5, 6-dimethylbenzimadazole) and this is also attached to corrin ring. The sixth group which makes up the octahedron is the active site, and is occupied by a CN^- group in cyanocobalamin. The CN^- is introduced in isolating the coenzyme, and is not present in the active form in living tissues. This position is occupied by OH^- in hydroxocobalamin, by water or by an organic group such as CH_3 (methylcobalamin) or adenosine. This shows that a metal to carbon σ bond can be formed. The cobalamins can be reduced from Co^{III} to Co^{II} and Co^{I} in neutral or alkaline solutions both in laboratory and in vivo (in the living body). The CoI complex is strongly reducing. Its structure is five-coordinate, *i.e.*, the site usually occupied by CN^- or OH^- is vacant.

Vitamin B_{12} is converted to coenzyme B_{12} by extracts from microorganisms supplemented with ATP Coenzyme B_{12} is associated with many biochemical reactions—

1. Conversion of methylmalonyl CoA to succinyl CoA.
2. Isomerisation of dicarboxylic acids, *e.g.*, glutamic acid into β-methyl-aspartic acid.
3. Dismutation of vicinal diols to the corresponding aldehydes, *e.g.*, propane-1, 2-diol into propionaldehyde.
4. Methylation of homocysteine to form methionine.

The Metals in I with the Appropriate Biomolecule in

	I	II
(A)	*Zn*	*(i) phosphotransferase*
(B)	*Cu*	*(ii) peroxidase*
(C)	*Fe*	*(iii) carboxypeptidase*
		(iv) haemocyanin
		(v) nitrogenase

Matched:

I	II	
(A) Zn	(i)	carboxypeptidase
(B) Cu	(iv)	haemocyanin (a copper containing protein)
(C) Fe	(ii)	peroxidase (a haeme enzyme) and
	(v)	nitrogenase (molybdenoferridoxin)

A few metalloproteins (enzymes) in which porphyrin ligands are involved. The structural features of the ligand systems, responsible for the enzyme activity.

The three major types of metalloproteins (enzyme) in which porphyrin ligands involved are *peroxidases, catalases* and *cytochrome* P_{450} enzymes.

Peroxidases catalyse the oxidation of a substrate by peroxides, mainly H_2O_2. Catalases catalyse decomposition of H_2O_2 (and some other peroxides) to H_2O and O_2. They have many similarities, both in structure and in aspects of their mechanisms. They both have high-spin ferric haeme groups lodged deeply in rather large protein molecules, with a histidine nitrogen atom constituting the fifth ligand. The sixth position may be occupied by a water molecule in the resting enzyme. There is growing evidence that a (proph) $Fe^{IV}=O$ species is a key intermediate in peroxidases and catalases.

The catalytic activity of peroxidase entails the following steps, where P represents the resting enzyme and AH a reducing substrate

$$P + H_2O_2 \longrightarrow \text{compound I} + H_2O$$
$$\text{compound I} + AH \longrightarrow \text{compound II} + A.$$
$$\text{compound II} + AH \longrightarrow P + A.$$

Compound I has an oxidation level two units higher than P, but it reverts to P in two one-electron steps. There is evidence to suggest that compound I and II can be more specifically formulated as follows—

$$\underset{P}{Fe^{III}} \xrightarrow{H_2O_2} \underset{I}{(R^{\cdot +})Fe^{IV}} \xrightarrow{AH} \underset{II}{Fe^{IV}—OH^{+}}$$

The $R^{\cdot +}$ represents a π-cation radical formed on the porphyrin ligand.

Cytochrome P_{450} enzymes are haeme containing oxygenases they catalyse the introduction of oxygen atoms from O_2 into substrates. Of the many possible substrates the most important are molecules in which C—H groups are converted to C—OH groups. The catalytic cycle entails a species in which the iron atom attains a high (IV or V) oxidation number. The coordination sphere of the iron atom includes, in addition to the porphyrins, one sulphur atoms, but whether the sixth position is occupied by water molecule or is vacant in the resting enzyme is still uncertain.

In step (1) of the catalytic cycle of Cytochrome P_{450} the substrate is bound, at or near the iron atom, and in step (2) reduction is affected by another enzymatic system. In step (3) the $(RH)Fe^{II}$ species binds an O_2 molecule. In step (4) another one electron reduction occur to give an iron (III) peroxo complex. In step (5), this complex loses H_2O to give the crucial high oxidation state $(RH)Fe^{IV}$=O.

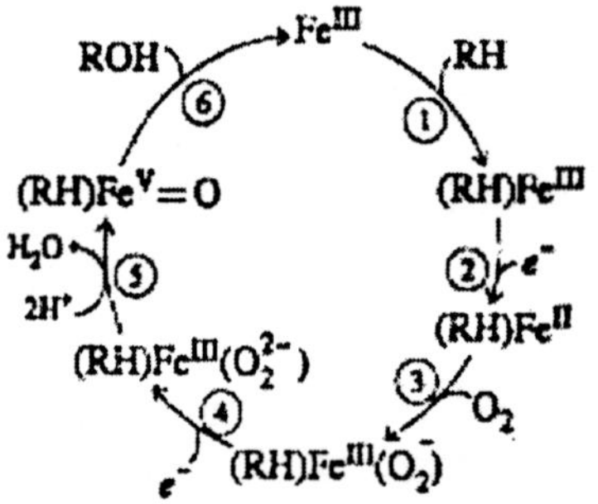

A cagtalytic cycle for cytochrome P_{450} enzyme.

Cytochrome c a Redox Protein: The active centre of Cytochrome *c* is the haeme group. It consists of a porphyrin ring chelated to an iron atom. The haeme group in Cytochrome *c* has a polypeptide chain attached and wrapped around it. This chain contains a variable number of amino acids ranging from 103 to 112. Å nitrogen atom from a histidine segment and a sulphur atom from a methionine segment of this chain are coordinated to the fifth and sixth coordination site. Thus, unlike the iron in haemoglobin and myoglobin there is no position for further coordination. Cytochrome *c* therefore cannot react by simple coordination but must react indirectly by an electron transfer mechanism. It can reduce the dioxygen and transmit its oxidising power towards the burning of food and release of energy in respiration.

The oxidation state of iron may be either + 2 or + 3 and the importance of the Cytochrome lies in their ability to act as redox intermediates in electron transfer.

Myoglobin is a protein of molecular weight of about 17,000 with the protein chain containing 153 amino acid residues folded about the single haeme group. This restricts access to the iron atom (by a second haeme) and reduces the likelihood of formation of a haematin-like Fe(III) dimer. The micro environment is similar to that in Cytochrome *c*, but there is no sixth ligand (methionine) to complete the coordination sphere of the iron atom. Thus there is a site to which a dioxygen molecule may reversibly bind.

Myoglobin contains iron (II) in the high spin state. The function of both haemoglobin and myoglobin is to bind oxygen, but their physiological roles are very different. Haemoglobin picks up oxygen in the lungs and carries it to tissues via, the circulatory system. Cellular oxygen is bound by myoglobin molecule that store it until it is required for metabolic action, whereupon they release it to other acceptors.

Role of Metal Complexes in Nitrogen Fixation: The discovery that dinitrogen was capable of forming stable complexes with transition metals led to extensive investigation of possibility of fixation via such complexes. For example, phosphine complexes of molybdenum and tungsten containing dinitrogen readily give ammonia in acidic media—

$$[MOCl_3(thf)_3] + 3e^- + 2N_2 + \text{excess dppe} \longrightarrow [Mo(N_2)_2\,(dppe)2] + 3Cl^-$$

$$[Mo(N2)2\,(dppe)\,2] + 6H^+ \longrightarrow 2NH_3 + N_2 + Mo^{IV}\text{ products.}$$

where thf = tetrahydrofuran and dppe = 1, 2-bis (diphenylphosphino) ethane, $Ph_2PCH_2CH_2PPh_2$. Both reactions takes place at room temperature and atmospheric pressure.

There are several bacteria and blue-green algae that can fix molecular nitrogen in vivo. Molybdenum is present in the catalysts of nitrogen fixing bacteria. The best known nitrogen fixing bacterium is Rhizobium. This contains the metalloenzyme nitrogenase. Nitrogenase contains two proteins, molybdoferredoxin and azoferredoxin. Molybdoferredoxin contains two Mo atoms, 24-36 Fe atom and 24-36 Sz atoms together with a protein and has a molecular weight of about 225000. Azoferredoxin is a derivative of ferredoxin $Fe_4S_4(SR)_4$ and has a molecular weight in the range of 50000-70000.

It is thought that N_2 bonds to Mo in the molybdoferredoxin. If the Mo is sufficiently reduce it can bond to antibonding orbitals on N_2. Then the Fe in azoferredoxin is reduced by free ferredoxin $Fe_4S_4(SR)_4$. An electron is transferred from reduced azoferredoxin to molybdoferredoxin, possibly via the Fe, then from the Mo to the N_2. Protons are then added to N_2, eventually giving NH_3. Adenosine triphosphate ATP is required to provide the necessary energy for the process.

Definition of Metalloenzymes

Metalloenzymes-Enzymes are large protein molecules so built that they can bind at least one reactant (substrate) and catalyse a biochemically important reaction. Some enzymes incorporate one or more metal atoms in their normal structure they are called metalloenzymes. The metal not merely participate during the time that the enzyme-substrate complex exists, but a permanent part of the enzyme. The metal atom or at least one of the metal atoms when two or more are present, occurs at or very near to the active site and plays a role in the activity of the enzyme. Ca, Mn, Fe, Cu, Zn and Mo have been found in metalloenzymes. Metalloenzymes are extremely efficient as catalyst, typically causing rates to increase 10^6 times or more.

Zinc Metalloenzymes: Two of the most important zinc metalloenzyme are the following:

Carbonic Anhydrase: (MW_2 = 30,000, 1 Zn) This enzyme occurs in red blood cells and catalyse the dehydration of bicarbonate ion and hydration of CO_2.

$$OH^- + CO_2 \rightleftharpoons HCO_3^-$$

In the absence of this enzyme the reaction is not compatible with physiological requirements.

Carboxy Peptidase: (MW = 34,300, 1 Zn) This enzyme in the pancreas of mammals catalyses the hydrolysis of the peptide bonded at the carboxyl end of a peptide chain.

Carrier of Natural Oxygen

The two most important metalloprotein, other than haemoglobin and myoglobin are haemerythrins and haemocyanin which act as natural oxygen carrier.

Haemerythrins: In a great variety of worms (marine) the oxygen-carrying molecules are iron bearing proteins, but they do not contain porphyrins, called haemerythrins. Its essential structural unit is a single chain protein of 113 amino acid residues, which contains two atoms of iron. Haemerythrins differ functionally from haemoglobin in two major ways. They show little or no cooperativity, and two iron atoms are employed to bind one O_2 molecule. Each dioxygen binding site contain two iron (II) atoms and the reaction takes place via a redox reaction to form iron (III) and peroxide (O_2^{2-}).

Structure of Ionophore

Ionophores (ion carriers) are lipophilic substances, capable of binding and carrying specific cations through the biological membranes. They differ from the uncouplers in that they promote the transport of cations other than H+ through the membrane.

The behaviour of valinomycin is typical of a group known as ionophore antibiotics. The antibiotic forms a lipid-soluble complex with K^+ which readily passes through the inner mitochondrial membrane, whereas K^+ alone in the absence of valinomycin penetrates only very slowly. Valinomycin binds K^+ more strongly than Na^+. Thus, valinomycin interferes with oxidative phosphorylation in mitochondria by making them permeable to K^+. The result is that mitochondria use the energy generated by electron transport to accumulate K^+ rather than to make ATP.

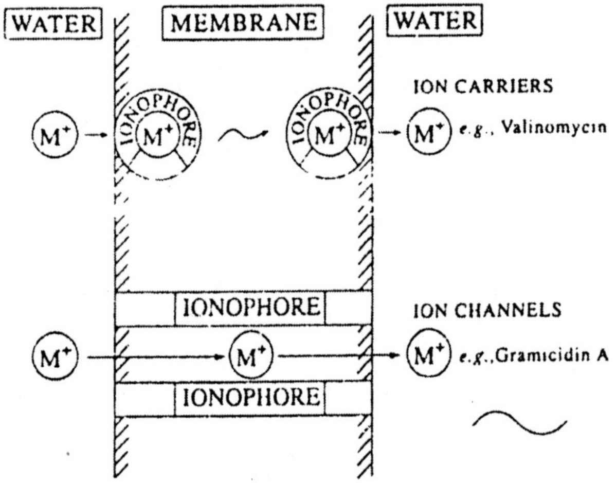

Schematic representation of the action of ionophores on membranes. M^+ = Metal ion

Molecular Mechanism of Ion Transport Across Membrane: Cells are enclosed by a membrane of about 70 Å thickness and composed of double layers of protein separated by lipids. Cation cannot pass through the lipid layer without encapsulation and thus the enclosed cation presents an organic, lipid soluble surface to the membrane.

Most animal cells have higher concentration of K^+ ions inside the cell membranes and a higher concentration of Na^+ outside the cell. Maintenance of these concentrations gradients requires a sodium and potassium pump. The immediate source of energy required to maintain this is ATP. In some cells, each

ATP molecule hydrolysed transports $3Na^+$ out of the cell and $2K^+$ into the cell. The K^+ is required in the cell for glucose metabolism, protein synthesis and activation of some enzymes. The transport of glucose and amino acids into the cell is coupled with Na^+ transport, which is favoured by great concentration gradient. The Na^+ entering the cell in this way must be pumped out again.

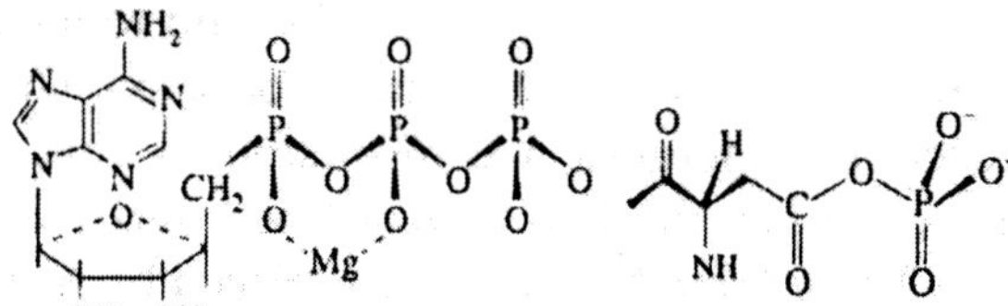

MgATP
I

β-Aspartyl phosphate
II

Using these clues, the following mechanism was proposed :

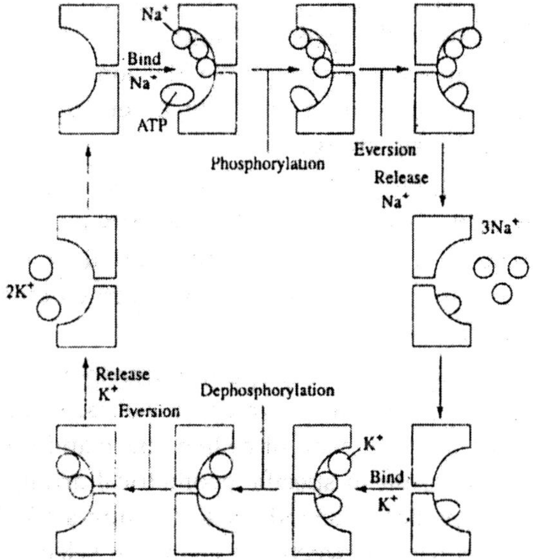

A schematic diagram of the mechanism of sodium pumping showing the role of eversion of the enzyme which transfers ions between the inside and outside of the cells.

A valuable clue to the manner in which iron pump functions, came in 1957 when an enzyme of molar mass 110 kg/mol was discovered by Jens Skou which hydrolyses ATP only if Na^+ and K^+ ions are present in addition to Mgt^+ required for all ATPases (enzymes catalysing the hydrolysis of ATP). The activity of this enzyme correlates quantitatively with the extent of ion transport. Another important clue was provided by the observation that this ATPase is phosphorylated at an aspartase site only in the presence of Na^+ and Mg^{2+} ions. The phosphorylated product is hydrolysed if K^+ ions are present. It has been also observed that the enzyme undergoes a conformational change when it is phosphorylated.

At first step ATP and three Na^+ ions bind to the inside of the membrane and the enzyme is phosphorylated. The product of the reaction undergoes a conformational change called eversion and brings the Na^+ ions to the outside of the cell membrane. There three Na^+ ions are replaced by two K^+ ions. The attachment of the K^+ ions induces dephosphorylation and hydrolysis of ATP to ADP produces a conformational change that carries two K^+ ions to the interior of the cell where they are released. The process builds up a charge gradient across the membrane because three Na^+ ions are released for two K^+ ions incorporated, and the outer surface becomes relatively positively charged.

The outstanding feature of the active transport is the selectivity of the complexation process, between K^+ and Na^+. The enzyme is selective first to Na^+ and then to K^+ in its different conformations. Selectivity is governed by coulombic forces and differences in ionic radii.

The ion channels which permit transport of ions, are constructed from protein polymers in helical conformations. These helices form the wall of cell channels and their precise conformations result in specific geometrical arrangements of a number of potential ligand-metal binding sites. There are striking differences between the channels, selective towards either Na^+ or K^+ ions.

Process of Photosynthesis

The photosynthetic process in green plants consists of splitting the elements of water, followed by reduction of carbon dioxide—

$$2H_2O \longrightarrow [4H] + O_2$$

$$x\,CO_2 + \frac{x}{2}\,[4H] \longrightarrow (CH_2O)_x + \frac{x}{2}O_2$$

where [4H] does not imply free atoms of hydrogen but a reducing capacity of four equivalents. In all dioxygen-producing organisms ranging from cyanobacteria to algae to higher plants, there are two coupled photosynthetic systems, PS-I and PS-II. The two differ in the type of chlorophyll present and in the accessory chemicals for processing the trapped energy of the photon. The primary product of PS-1 is reduced carbon, and the primary product of PS-II is energy in the form to two moles of ATP with molecular oxygen as a chemical by product.

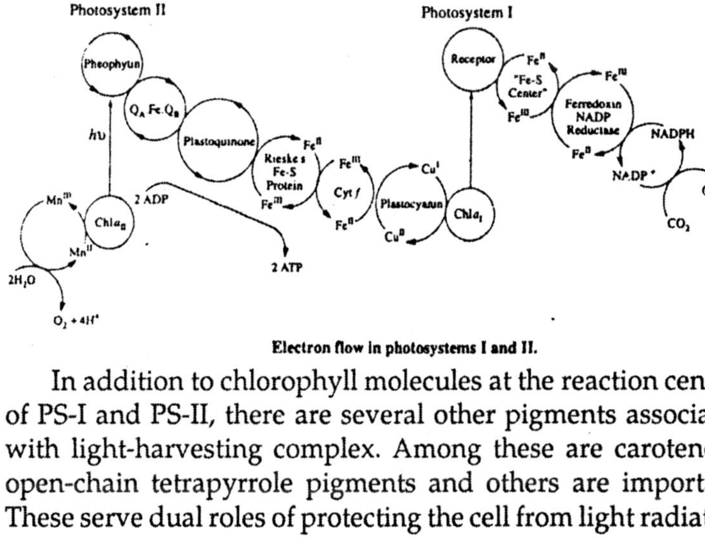

Electron flow in photosystems I and II.

In addition to chlorophyll molecules at the reaction centres of PS-I and PS-II, there are several other pigments associated with light-harvesting complex. Among these are carotenoid, open-chain tetrapyrrole pigments and others are important. These serve dual roles of protecting the cell from light radiation and at the same time harvesting much of it for photosynthesis. Some of these compounds are arranged in antenna-like rods that gather the light energy and funnel it into the reaction centres.

The energy of an absorbed photon in either PS-I or PS-II initiates a series of redox reactions. PS-I produces a moderately strong reducing species (RED_I) and a moderately strong oxidising species (OX_I). PS-II provides a stronger oxidising agent (OX_{II}) but a weaker reducing agent (RED_{II}).

OX_{II} is responsible for the production of molecular oxygen in photosynthesis. Å manganese complex, probably with four atoms of manganese, is attached to a protein molecule. It reduces OX_{II} which is recycled for use by another excited chlorophyll molecule in PS-II. In the redox reaction the manganese shuttles between two oxidation states with each manganese atom increasing its oxidation state by one unit, but it is not known with certainty what these oxidation are [three Mn(II) and one Mn(III) or three Mn(III) and one Mn(IV)]. Butterfly cluster of Mn_4 configuration have been suggested along with others.

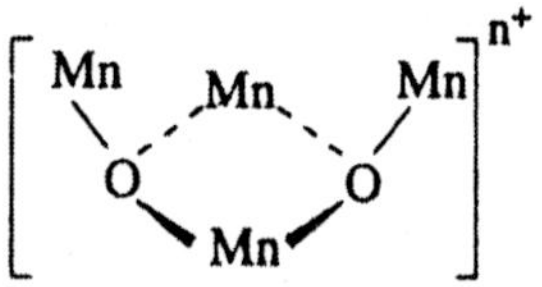

Copper Metalloenzymes and Metalloprotein: The copper enzymes are mostly oxidases, that is enzymes which catalyse oxidations. Examples are:

Ascorbic acid oxidase: (MW = 1,40,000; 8 Cu). It is widely distributed in plants and micro-organisms. It catalyses oxidation of ascorbic acid (vitamin C) to dehydroascorbic acid.

Cytochrome Oxidase: The terminal electron acceptor in the oxidative pathway of cell mitochondria. This enzyme also contains haeme.

Tyrosinase: Various tyrosinase catalyse the formation of pigments (melanins) in a host of plants and animals.

Copper Metalloprotein (Haemocyanin): In the phyla Mollusca and Arthropoda the oxygen carrying molecule is a copper containing protein is haemocyanin. They are very large of very high molecular weight, but they are consist of subunits. Each subunit contains a pair of Cu atoms, and they can bind one O_2 molecule per pair of Cu atoms. It generally display cooperativity in O_2 binding.

7

Petrochemical Effects

An oil refinery is an industrial process plant where crude oil is processed and refined into more useful petroleum products, such as gasoline, diesel fuel, asphalt base, heating oil, kerosene, and liquefied petroleum gas. Oil refineries are typically large sprawling industrial complexes with extensive piping running throughout, carrying streams of fluids between large chemical processing units.

Various Chemicals

Ethanol in the past has been used commercially to synthesise dozens of other high-volume chemical commodities. At the present, it has been supplanted in many applications by less costly petrochemical feedstocks.

However, in markets with abundant agricultural products, but a less developed petrochemical infrastructure, such as the People's Republic of China, Pakistan, India, and Brazil, ethanol can be used to produce chemicals that would be produced from petroleum in the West, including ethylene and butadiene.

Other Uses

Ethanol is easily soluble in water in all proportions with a slight overall decrease in volume when the two are mixed. Absolute ethanol and 95 per cent ethanol are themselves good solvents, somewhat less polar than water and used in perfumes, paints and tinctures. Other proportions of ethanol with water or other solvents can also be used as a solvent. Alcoholic drinks have a large variety of tastes because various flavour compounds are dissolved during brewing. When ethanol is produced as a mixing beverage, it is a neutral grain spirit.

Ethanol is used in medical wipes and in most common antibacterial hand sanitiser gels at a concentration of about 62 per cent (percentage by weight, not volume) as an antiseptic. The peak of the disinfecting power occurs around 70 per cent ethanol; stronger and weaker solutions of ethanol have a lessened ability to disinfect. Solutions of this strength are often used in laboratories for disinfecting work surfaces. Ethanol kills organisms by denaturing their proteins and dissolving their lipids and is effective against most bacteria and fungi, and many viruses, but is ineffective against bacterial spores. Alcohol does not act like an antibiotic and is not effective against infections by ingestion. Ethanol in the low concentrations typically found in most alcoholic beverages does not have useful disinfectant or antiseptic properties, internally or externally. Ethanol is often used as an antidote in cases of methanol poisoning.

Meaning and Uses of Petrochemicals

Petrochemicals are chemical products made from raw materials of petroleum (hydrocarbon) origin. Although some of the chemical compounds which originate from petroleum may also be derived from other sources such as coal or natural gas, petroleum is a major source of many. The two main classes of petrochemical raw materials are olefins (including ethylene and propylene) and aromatics (including benzene and xylene isomers), both of which are produced in very large quantities. At oil refineries, olefins are produced mainly from hydrocarbons by chemical cracking such as steam cracking and by catalytic reforming. At oil refineries, aromatic hydrocarbons are mainly produced by catalytic reforming or similar processes.

From these basic building blocks are made a very wide range of chemicals and other materials used in industry—monomers, solvents, detergents, adhesives, etc. From the monomers, polymers or oligomers are produced for plastics, resins, fibres, elastomers, certain lubricants and gels.

World is production of ethylene is around 110 million tons per annum, of propylene 65 million tons, and of aromatic raw materials 70 million tons. The largest petrochemical industries are to be found in the USA and Western Europe, though the major growth in new production capacity is in the Middle East and Asia. There is a substantial inter-regional trade in petrochemicals of all kinds.

The following is a partial list of the major commercial petrochemicals and their derivatives:

- *Ethylene:* The simplest olefin; used as a ripening hormone, a monomer and a chemical feedstock
  - — *Polyethylenes:* Polymerised ethylene
  - — *Ethanol:* Made by hydration (chemical reaction adding water) of ethylene
  - — *Ethylene Oxide:* sometimes called oxirane; can be made by oxidation of ethylene
    - ⇒ *Ethylene Glycol:* From hydration of ethylene oxide or oxidation of ethylene
    - ⇒ *Engine Coolant:* Contains ethylene glycol
    - ⇒ *Polyesters:* Any of several polymers with ester linkages in the backbone chain
    - ⇒ Glycol Ethers: From condensation of glycols
    - ⇒ Ethoxylates

- Vinyl acetate
- 1,2-dichloroethane
    - ⇒ Trichloroethylene
    - ⇒ *Tetrachloroethylene:* Also called perchloroethylene; used as a dry cleaning solvent and degreaser
    - ⇒ *Vinyl Chloride:* Monomer for polyvinyl chloride
    - ⇒ *Polyvinyl Chloride (PVC):* Type of plastic used for piping, tubing, other things

• *Propylene:* Used as a monomer and a chemical feedstock

- Isopropyl alcohol - 2-propanol; often used as a solvent or rubbing alcohol
- *Acrylonitrile:* Useful as a monomer in forming Orlon, ABS
- *Polypropylene:* Polymerised propylene
- Propylene oxide
    - ⇒ *Propylene Glycol:* Sometimes used in engine coolant
    - ⇒ *Glycol Ethers:* From condensation of glycols
- *Isomers of Butylene:* Useful as monomers or co-monomers
    - ⇒ *Isobutylene:* Feed for making [[methyl *tert*-butyl ether]] (MTBE) or monomer for copolymerisation with a low percentage of isoprene to make butyl rubber
- *1,3-butadiene:* A diene often used as a monomer or co-monomer for polymerisation to elastomers such as polybutadiene or a plastic such as acrylonitrile-butadiene-styrene (ABS)
    - ⇒ *Synthetic Rubbers:* Synthetic elastomers made of any one or more of several petrochemical (usually) monomers such as 1, 3-butadiene,

styrene, isobutylene, isoprene, chloroprene, etc. Elastomeric polymers are often made with a high percentage of conjugated diene monomers such as 1, 3-butadiene, isoprene, or chloroprene.

— Higher olefins

⇒ Polyolefins such poly-alpha-olefins which are used as lubricants.

⇒ *Alpha-olefins:* Used as monomers, co-monomers, and other chemical precursors. For example, a small amount of 1-hexene can be copolymerised with ethylene into a more flexible form of polyethylene.

⇒ Other higher olefins

⇒ Detergent alcohols

— Acrylic acid

⇒ Acrylic polymers

— Allyl chloride

⇒ *Epichlorohydrin:* Chloro-oxirane; used in epoxy resin formation.

⇒ *Epoxy resins:* A type of polymerising glue from bisphenol A, epichlorohydrin, and some amine.

- *Benzene:* The simplest aromatic hydrocarbon

— *Ethylbenzene:* Made from benzene and ethylene

⇒ Styrene made by dehydrogenation of ethylbenzene; used as a monomer.

⇒ *Polystyrenes:* Polymers with styrene as a monomer.

— *Cumene:* Isopropylbenzene; a feedstock in the cumene process

⇒ *Phenol:* Hydroxybenzene; often made by the cumene process.

⇒ *Acetone:* Dimethyl ketone; also often made by the cumene process.

⇒ *Bisphenol A:* A type of "double" phenol used in polymerisation in epoxy resins and making a common type of polycarbonate.

⇒ *Epoxy Resins:* A type of polymerising glue from bisphenol A, epichlorohydrin, and some amine.

⇒ *Polycarbonate:* A plastic polymer made from bisphenol A and phosgene (carbonyl dichloride).

⇒ *Solvents:* Liquids used for dissolving materials. Examples often made from petrochemicals include ethanol, isopropyl alcohol, acetone, benzene, toluene, xylenes.

— *Cyclohexane:* A 6-carbon aliphatic cyclic hydrocarbon sometimes used as a non-polar solvent

⇒ *Adipic Acid:* A 6-carbon dicarboxylic acid which can be a precursor used as a co-monomer together with a diamine to form an alternating copolymer form of nylon.

⇒ *Nylons:* Types of polyamides, some are alternating copolymers formed from copolymerising dicarboxylic acid or derivatives with diamines.

⇒ *Caprolactam:* A 6-carbon cyclic amide.

— *Nitrobenzene:* Can be made by single nitration of benzene

⇒ *Aniline:* aminobenzene

⇒ *Methylene Diphenyl Di-isocyanate (MDI):* Used as a co-monomer with diols or polyols to form polyurethanes or with di- or polyamines to form polyureas.

⇒ Polyurethanes

— *Alkylbenzene:* A general type of aromatic hydrocarbon which can be used as a precursor for a sulphonate surfactant (detergent)

⇒ *Detergents:* Often include surfactants types such as alkylbenzenesulphonates and nonylphenol ethoxylates.

— Chlorobenzene

Toluene: Methylbenzene; can be a solvent or precursor for other chemicals

— Benzene

— *Toluene Di-isocyanate (TDI):* Used as co-monomers with diols or polyols to form polyurethanes or with di- or polyamines to form polyureas

⇒ Polyurethanes: A polymer formed from diisocyanates and diols or polyols

— Benzoic Acid: Carboxybenzene

⇒ Caprolactam

⇒ Nylon

- *Mixed Xylenes:* Any of three dimethylbenzene isomers, could be a solvent but more often precursor chemicals

— *Ortho-xylene:* Both methyl groups can be oxidised to form ortho phthalic acid

⇒ Phthalic anhydride

— *Para-xylene:* Both methyl groups can be oxidised to form terephthalic acid

⇒ *Dimethyl Terephthalate:* Can be copolymerised to form certain polyesters.

⇒ *Polyesters:* Although there can be many types, polyethylene terephthalate is made from petrochemical products and is very widely used.

⇒ *Purified Terephthalic Acid:* Often copolymerised to form polyethylene terephthalate.

⇒ Polyesters

Importance of Oil Refinery

Raw oil or unprocessed ("crude") oil is not very useful in the form it comes in out of the ground. Although "light, sweet" (low viscosity, low sulphur) oil has been used directly as a burner fuel for steam vessel propulsion, the lighter elements form explosive vapours in the fuel tanks and so it is quite dangerous, especially so in warships.

For this and many other uses, the oil needs to be separated into parts and refined before use in fuels and lubricants, and before some of the by-products could be used in petrochemical processes to form materials such as plastics, and foams. Petroleum fossil fuels are used in ship, automobile and aircraft engines. These different hydrocarbons have different boiling points, which means they can be separated by distillation. Since the lighter liquid elements are in great demand for use in internal combustion engines, a modern refinery will convert heavy hydrocarbons and lighter gaseous elements into these higher value products using complex and energy intensive processes.

Oil can be used in so many various ways because it contains hydrocarbons of varying molecular masses, forms and lengths such as paraffins, aromatics, naphthenes (or cycloalkanes), alkenes, dienes, and alkynes.

Hydrocarbons are molecules of varying length and complexity made of only hydrogen and carbon atoms. Their various structures give them their differing properties and thereby uses. The trick in the oil refinement process is separating and purifying these.

Once separated and purified of any contaminants and impurities, the fuel or lubricant can be sold without any further processing. Smaller molecules such as isobutane and propylene or butylenes can be recombined to meet specific octane requirements of fuels by processes such as alkylation or less commonly, dimerisation.

Octane grade of gasoline can also be improved by catalytic reforming, which strips hydrogen out of hydrocarbons to produce aromatics, which have much higher octane ratings. Intermediate products such as gasoils can even be reprocessed to break a heavy, long-chained oil into a lighter short-chained one, by various forms of cracking such as Fluid Catalytic Cracking, Thermal Cracking, and Hydrocracking. The final step in gasoline production is the blending of fuels with different octane ratings, vapour pressures, and other properties to meet product specifications.

Oil refineries are large scale plants, processing from about a hundred thousand to several hundred thousand barrels of crude oil per day. Because of the high capacity, many of the units are operated continuously (as opposed to processing in batches) at steady state or approximately steady state for long periods of time (months to years).

Major Products of Oil Refineries

- Asphalt
- Diesel fuel
- Naphtha
- Fusel oils
- Gasoline
- Kerosene and related jet aircraft fuels
- Liquid petroleum gas (LPG)
- Lubricating oils
- Paraffin wax
- Tar

Common Process Units Found in a Refinery

- Desalter Unit (washes out salt from the crude oil before it goes into the atmospheric distillation unit).
- Atmospheric Distillation Unit (distills crude oil into fractions).
- Vacuum Distillation Unit (further distills residual bottoms after atmospheric distillation).
- Naphtha Hydrotreater Unit (uses hydrogen to desulphurise naphtha from atmospheric distillation. Must hydrotreat the naphtha before sending to a Catalytic Reformer Unit.).
- Catalytic Reformer Unit (contains catalyst used to convert the naphtha-boiling range molecules into higher octane reformate (reformer product). The reformate has higher content of aromatics, olefins, and cyclic hydrocarbons). An important by-product of a reformer is hydrogen released during the catalyst reaction. The hydrogen is used either in the hydrotreaters and hydrocracker.).
- Distillate Hydrotreater Unit (desulphurises distillate (diesel) after atmospheric distillation).
- Fluid Catalytic Cracking (FCC) Unit (upgrades heavier fractions into lighter, more valuable products).
- Hydrocracker Unit (uses hydrogen to upgrade heavier fractions into lighter, more valuable products).
- *Merox Treater:* In certain cases such as for jet fuel or kerosene treatment, a Merox process unit may be used to oxidise mercaptans to organic disulphides.
- Coking unit (processes asphalt into gasoline and diesel fuel, leaving coke as a residual product).
- Alkylation unit (produces high octane component for gasoline blending).
- Dimerisation Unit.

- Isomerisation Unit (converts linear molecules to higher octane branched molecules for blending into gasoline or feed to alkylation units).
- Steam reforming Unit (produces hydrogen for the hydrotreaters or hydrocracker).
- Liquified gas storage units for propane and similar gaseous fuels at pressure sufficient to maintain in liquid form - these are usually spherical or bullets (horizontal cylinder with rounded ends).
- Storage tanks for crude oil and finished products, usually cylindrical, with some sort of vapour enclosure and surrounded by an earth berm to contain spills.
- Amine gas treater, Claus unit, and tail gas treatment for processing hydrogen sulphide from hydrodesulphurisation.
- Utility units such as cooling towers for circulating cooling water, boiler plants for steam generation, and wastewater collection and treating systems to make such water suitable for reuse or for disposal.

Speciality end Products

These will blend various feedstocks, mix appropriate additives, provide short term storage, and prepare for bulk loading to trucks, barges, product ships, and railcars.

- Gaseous fuels such as propane, stored and shipped in liquid form under pressure in specialised railcars to distributors.
- Liquid fuels blending (producing automotive and aviation grades of gasoline, kerosene, various aviation turbine fuels, and diesel fuels, adding dyes, detergents, antiknock additives, oxygenates, and antifungal compounds as required). Shipped by barge, rail, and tanker ship. May be shipped regionally in dedicated pipelines to point consumers, particularly aviation jet

fuel to major airports, or piped to distributors in multi-product pipelines using product separators called pipeline inspection gauges ("pigs").

- Lubricants (produces light machine oils, motor oils, and greases, adding viscosity stabilisers as required), usually shipped in bulk to an offsite packaging plant.
- Wax (paraffin), used in the packaging of frozen foods, among others. May be shipped in bulk to a site to prepare as packaged blocks.
- Sulphur (or sulphuric acid), by-products of sulphur removal from petroleum which may have up to a couple per cent sulphur as organic sulphur-containing compounds. Sulphur and sulphuric acid are useful industrial materials. Sulphuric acid is usually prepared and shipped as the acid precursor oleum.
- Bulk tar shipping for offsite unit packaging for use in tar-and-gravel roofing.
- Asphalt unit. Prepares bulk asphalt for shipment.
- Petroleum coke, used in speciality carbon products or as solid fuel.
- Petrochemicals or petrochemical feedstocks, which are often sent to petrochemical plants for further processing in a variety of ways. The petrochemicals may be olefins or their precursors, or various types of aromatic petrochemicals.

Co-plant Siting

Frequently a chemical plant will be sited adjacent to a refinery, utilising intermediate products as feedstocks for the production of specialised materials such as plastics or agrichemicals.

Safety and Environmental Concerns

The refining process releases numerous different chemicals into the atmosphere; consequently, there are substantial air pollution emissions and a notable odour normally accompanies the presence of a refinery. Aside from air pollution impacts, there are also wastewater concerns, upset risks of fire and explosion, and both occupational noise and environmental noise health effects.

The public has demanded that many governments place restrictions on contaminants that refineries release, and most refineries have installed the equipment needed to comply with the requirements of the pertinent environmental protection regulatory agencies. In the United States, there is strong pressure to prevent the development of new refineries, and no major refinery has been built in the country since Marathon's Garyville, Louisiana facility in 1976. However, many existing refineries have been expanded during that time. Environmental restrictions and pressure to prevent construction of new refineries may have also contributed to rising fuel prices in the United States. Additionally, many refineries (over 100 since the 1980s) have closed due to obsolescence and/or merger activity within the industry itself. This activity has been reported to Congress and in specialised studies not widely publicised.

Environmental and safety concerns mean that oil refineries are sometimes located some distance away from major urban areas. Nevertheless, there are many instances where refinery operations are close to populated areas and pose health risks such as in the Campo de Gibraltar, a Spanish state owned refinery near the towns of Gibraltar, Algeciras, La Linea, San Roque and Los Barrios with a combined population of over 300,000 residents within a 5 mile radius and the CEPSA refinery in Santa Cruz on the island of Tenerife, Spain which is sited in a densely-populated city centre and next to the only two major evacuation routes in and out of the city.

In California's Contra Costa County and Solano County, a shoreline necklace of refineries and associated chemical plants are adjacent to urban areas in Richmond, Martinez, California,

Pacheco, Concord, Pittsburg, Vallejo and Benicia, with occasional accidental events that require "shelter in place" orders to the adjacent populations.

History

The world's first oil refineries were set up by Ignacy Lukasiewicz near Jaslo, Austrian Empire (now in Poland) in the years 1854-56 but they were initially small as there was no real demand for refined fuel. As Lukasiewicz's kerosene lamp gained popularity the refining industry grew in the area.

The first large oil refinery opened at Ploiesti, Romania in 1856. Several other refineries were built at that location with investment from United States companies before being taken over by Nazi Germany during World War II. Most of these refineries were bombarded by US Army Air Forces in Operation Tidal Wave, August 1, 1943. Since then they have been rebuilt, and currently pose somewhat of an environmental concern.

Another early example is Oljeon, now preserved as a museum at the UNESCO world heritage site Engelsberg. It started operation in 1875 and is part of the Ecomuseum Bergslagen.

It is difficult to exactly state which is largest oil refinery in the world. At one time this was claimed to be Ras Tanura, Saudi Arabia, owned by Saudi Aramco. For most of the 20th century, the largest refinery of the world was that of Abadan in Iran. This refinery suffered extensive damage during the Iran-Iraq war. The Guinness Book of World Records now (October 2006) records the BP Amoco refinery in Texas City, USA as the refinery with the largest capacity (433,000 barrels/day which is approximately 15,100,000 Imperial gallons/day or 68,800,000 litres/day). Early US refineries processed crude oil to recover the kerosene.

Other products (like gasoline) were considered wastes and were often dumped directly into the nearest river. The invention of the automobile shifted the demand to gasoline and diesel, which remain the primary refined products today. Refineries predating the EPA were very toxic to the environment. Strict legislation has mandated that refineries meet modern air and water cleanliness standards. In fact, obtaining a permit to build even a modern refinery with minimal impact on the environment (other than CO_2 emissions) is so difficult and costly that no new refineries have been built (though many have been expanded) in the United States since 1976. As a result, some believe that this may be the reason that the US is becoming more and more dependent on the imports of finished gasoline, as opposed to incremental crude oil. On the other hand, studies have revealed that accelerating merger activity in the refining and production sector has reduced capacity further, resulting in tighter markets in the United States in particular.

Forms of Monomers

A monomer (from Greek mono "one" and meros "part") is a small molecule that may become chemically bonded to other monomers to form a polymer.

Examples of monomers are hydrocarbons such as the alkene and arene homologous series. Here hydrocarbon monomers such as phenylethene and ethene form polymers used as plastics like polyphenylethene (commonly known as polystyrene) and polyethene (commonly known as polyethylene or polythene). Other commercially important monomers include acrylic monomers such as acrylic acid, methyl methacrylate, and acrylamide.

Amino acids are natural monomers, and polymerise to form proteins. Glucose monomers can also polymerise to form starches, amylopectins and glycogen polymers. In this case, the polymerisation reaction is known as a dehydration or condensation reaction (due to the formation of water (H_2O) as one of the products) where a hydrogen atom and a hydroxyl (-OH) group are lost to form H_2O and an oxygen molecule bonds between each monomer unit.

The lower molecular weight compounds built from monomers are also referred to as dimers, trimers, tetramers, quadramers, pentamers, octamers, 20-mers, etc. if they have 2, 3, 4, 5, 8, or 20 monomer units, respectively. Any number of these monomer units may be indicated by the appropriate prefix, e.g., *deca*mer, being a 10-unit monomer chain or polymer. Larger numbers are often stated in English in lieu of Greek. Polymers with relatively low number of units are called oligomers.

Fundamentals of Ethylene

Ethylene (or IUPAC name ethene) is the chemical compound with the formula C_2H_4. It is the simplest alkene. Because it contains a double bond, ethylene is called an unsaturated hydrocarbon or an olefin. It is exceedingly important in industry and even has a role in biology as a hormone. Ethylene is the organic compound produced on the largest scale: global production of ethylene exceeded 75 million metric tons per year in 2005.

This hydrocarbon has four hydrogen atoms bound to a pair of carbon atoms that are connected by a double bond. All six atoms that comprise ethylene are coplanar. The H-C-H angle is 117°, close to the 120°F or ideal sp^2 hybridised carbon. The molecule is also relatively rigid: rotation about the C-C bond is a high energy process that requires breaking the σ-bond, while retaining the σ-bond between the carbon atoms.

The double bond is a region of high electron density, and most reactions occur at this double bond position.

History

From 1795 on, ethylene was referred to as the *olefiant gas* (oil-making gas), because it combined with chlorine to produce the *oil of the Dutch chemists* (1, 2-dichloroethane), first synthesised in 1795 by a collaboration of four Dutch chemists.

In the mid-19th century, the suffix *-ene* (an Ancient Greek root added to the end of female names meaning "daughter of") was widely used to refer to a molecule or part thereof that contained one fewer hydrogen atoms than the molecule being modified. Thus, *ethylene* (C_2H_4) was the "daughter of ethyl" (C_2H_5). The name ethylene was used in this sense as early as 1852.

In 1866, the German chemist, Augustus von Hofmann proposed a system of hydrocarbon nomenclature in which the suffixes -ane, -ene, -ine, -one, and -une were used to denote the hydrocarbons with 0, 2, 4, 6, and 8 fewer hydrogens than their parent alkane. In this system, ethylene became *ethene*. Hofmann's system eventually became the basis for the Geneva nomenclature approved by the International Congress of Chemists in 1892, which remains at the core of the IUPAC nomenclature. However, by that time, the name ethylene was deeply entrenched, and it remains in wide use today, especially in the chemical industry.

The 1979 IUPAC nomenclature rules make an exception for retaining the non-systematic name ethylene, however, this decision was reversed in the 1993 rules.

Production

Ethylene is produced in the petrochemical industry by steam "cracking". In this process, gaseous or light liquid hydrocarbons are briefly heated to 750-950°C, inducing numerous free radical reactions.

This process converts large hydrocarbons into smaller ones and introduces unsaturation. Ethylene is separated from the resulting complex mixture by repeated compression and distillation. In a related process used in oil refineries, high molecular weight hydrocarbons are cracked over Zeolite catalysts. Heavier feedstocks, such as naphtha and gas oils require at least two "quench towers" downstream of the cracking furnaces to recirculate pyrolysis-derived gasoline and process water. When cracking a mixture of ethane and propane, only one water quench tower is required.

The areas of an ethylene plant are: (1) steam cracking furnaces; (2) primary and secondary heat recovery with quench; (3) a dilution steam recycle system between the furnaces and the quench system; (4) primary compression of the cracked gas (3 stages of compression); (5) hydrogen sulphide and carbon dioxide removal (acid gas removal); (6) secondary compression (1 or 2 stages); (7) drying of the cracked gas; (8) cryogenic treatment; (9) all of the cold cracked gas stream goes to the demethaniser tower. The overhead stream from the demethaniser tower consists of all the hydrogen and methane that was in the cracked gas stream. Different methods of cryogenically treating this overhead stream results in the separation of the hydrogen and the methane. This usually involves liquid methane at a temperature around 250°F. Complete recovery of all the methane is critical to the economical operation of an ethylene plant. (10) the bottom stream from the demethaniser tower goes to the deethaniser tower. The overhead stream from the deethaniser tower consists of all the C_2's that were in the cracked gas stream.

The C_2's then go to a C_2 splitter. The product ethylene is taken from the overhead of the tower and the ethane coming from the bottom of the splitter is recycled to the furnaces to be cracked again; (11) the bottom stream from the deethaniser tower goes to the depropaniser tower. The overhead stream from the depropaniser tower consists of all the C_3's that were in the cracked gas stream. Prior to sending the C_3's to the C_3 splitter this stream is hydrogenated in order to react out the methylacetylene and propadiene. Then this stream is sent to the C_3 splitter. The overhead stream from the C_3 splitter is product propylene and the bottom stream from the C_3 splitter is propane which can be sent back to the furnaces for cracking or used as fuel. (12) The bottom stream from the depropaniser tower is fed to the debutaniser tower. The overhead stream from the debutaniser is all of the C_4's that was in the cracked gas stream. The bottom stream from the debutaniser consists of everything in the cracked gas stream that is C_5 or heavier. This could be called a light pyrolysis gasoline.

Since the production of ethylene is energy intensive, much effort has been dedicated recovering heat from the gas leaving the furnaces. Most of the energy recovered from the cracked gas is used to make high pressure (1200 psig) steam. This steam is in turn used to drive the turbines for compressing cracked gas, the propylene refrigeration compressor, and the ethylene refrigeration compressor. An ethylene plant, once running, does not need to import any steam to drive its' steam turbines. A typical world scale ethylene plant (about 1.5 billion pounds of ethylene per year) uses a 45,000 horsepower cracked gas compressor, a 30,000 horsepower propylene compressor, and a 15,000 horsepower ethylene compressor.

When starting an ethylene plant, it is important to start the cooling systems in the proper order. The cooling systems consist of Cooling Tower Water (CTW); propylene refrigeration with four or five different levels or stages. Each level corresponds to a particular pressure and temperature; and three or four stages of ethylene regfrigeration. The CTW must be started first because the propylene system needs it to condense propylene and the ethylene refrigeration systems needs it to desuperheat high pressure ethylene.

The propylene system must start next because the ethylene system needs high pressure propylene for desuperheating the high pressure ethylene stage and the low pressure propylene stage for condensing the high pressure ethylene. While the ethylene plant is running, the plant can continue to run for a time if the ethylene refrigeration compressor shuts down. However, if the propylene compressor shuts down the whole plant must be shut down immediately.

Theoretical Considerations

Although ethylene is a relatively simple molecule, its spectrum is considered to be one of the most difficult to explain adequately from both a theoretical and practical perspective. For this reason, it is often used as a test case in computational chemistry. Of particular note is the difficulty in characterising the ultraviolet absorption of the molecule. Interest in the subtleties and details of the ethylene spectrum can be dated back to at least the 1950s.

Chemical Reactions

Ethylene is an extremely important building block in the petrochemical industry. It can undergo many types of reactions which leads to a plethora of major chemical products. A list of some major types of reactions includes, (1) Polymerisation, (2) Oxidation, (3) Halogenation and Hydrohalogenation, (4) Alkylation, (5) Hydration, (6) Oligomerisation, (7) Oxo-reaction, and (8) a ripening agent for fruits and vegetables.

Additions to Double Bond

Like most alkenes, ethylene reacts with halogens to produce halogenated hydrocarbons1,2-$C_2H_4X_2$. It can also react with water to produce ethanol, but the rate at which this happens is very slow unless a suitable catalyst, such as phosphoric or sulphuric acid, is used. Under high pressure, and, in the presence of a catalytic metal (platinum, rhodium, nickel), hydrogen will react with ethylene.

Ethylene is used primarily as an intermediate in the manufacture of other chemicals used in the synthesis of monomers. Ethylene can be chlorinated to produce 1,2-dichloroethane (ethylene dichloride), which can be converted to vinyl chloride, the monomer precursor to plastic polyvinyl chloride, or combined with benzene to produce ethylbenzene, which is used in the manufacture of polystyrene, another important plastic.

Ethylene is more reactive than any alkanes because of two reasons:

1. It has a double bond, one called the π-bond(pi) and one called the σ-bond (sigma). Where π-bond is weak and σ-bond is strong. The presence of the π-bond makes it a high energy molecule. Thus bromine water decolourises readily.
2. High electron density at the double bond makes it react readily. It is broken in an addition reaction to produce many useful products.

Polymerisation

Ethylene polymerises to produce polyethylene, also called *polyethene* or *polythene*, the world's most widely-used plastic.

Major polyethylene product groups are low density polyethylene, high density polyethylene, polyethylene copolymers, as well as ethylene-propylene co- and terpolymers.

Oxidation

Ethylene is oxidised to produce ethylene oxide, which is hydrolysed to ethylene glycol. It is also a precursor to vinyl acetate.

Ethylene undergoes oxidation by palladium to give acetaldehyde. This conversion was at one time a major industrial process. The process proceeds via the initial complexation of ethylene to a Pd (II) centre.

Major intermediates of the oxidation of Ethylene are ethylene oxide, acetaldehyde, vinyl acetate and ethylene glycol. The list of products made from these intermediates is long. Some of them are: polyesters, polyurethane, morpholine, ethanolamines, aspirin and glycol ethers.

Halogenation and Hydrohalogenation

Major intermediates from the halogenation and hydrohalogenation of ethylene include: ethylene dichloride, ethyl chloride and ethylene dibromide. Some products in this group are: polyvinyl chloride, trichloroethylene, perchloroethylene, methyl chloroform, polyvinylidiene chloride and copolymers, and ethyl bromide.

Alkylation

Major chemical intermediates from the alkylation of ethylene include: ethylbenzene, ethyl toluene, ethyl anilines, 1,4 hexadiene and aluminium alkyls. Products of these intermediates include polystyrene, unsaturated polyesters and ethylene-propylene terpolymers.

Hydration

Ethanol is the primary intermediate of the hydration of ethylene. Important products from ethanol are: ethylamines, yeast, acetaldehyde, and ethyl acetate.

Oligomerisation

The primary products of the Oligomerisation of ethylene are alpha-olefins and linear primary alcohols. These are used as plasticisers and surfactants.

Oxo-reaction

The Oxo-reaction of ethylene results in propionaldehyde with its' primary products of propionic acid and n-propyl alcohol.

In the Synthesis of Fine Chemicals

Ethylene is useful in organic synthesis. Representative reactions include Diels-Alder additions, ene reaction, and arene alkylation.

Miscellaneous

Ethylene was once used as a general anesthetic applicable via inhalation, but it has long since been replaced.

It has also been hypothesised that ethylene was the catalyst for utterances of the oracle at Delphi in ancient Greece.

It is also found in many lip gloss products. Production of Ethylene in mineral oil filled transformers is a key indicator of severe localised overheating (>750°C.)

Process of Antifreeze

Antifreeze is used in internal combustion engines, and for many other heat transfer applications, such as electronics cooling and chillers for HVAC. Compounds are added to water to reduce the freezing point of the mixture to below the lowest temperature that the system is likely to be exposed to, and to inhibit corrosion in cooling systems which often contain a range of electrochemically incompatible metals (aluminium, cast iron, copper, lead solder, etc.). The term 'colligative agent' is to be preferred as, in warm climates, the benefit of these compounds is to increase the boiling point of the coolant, which should then be more properly referred to as 'anti-boil', and as antifreeze decreases and increases both properties, respectively, 'colligative agent' more accurately describes the liquid. The term 'engine coolant' is widely used in industry.

Agents of Antifreeze

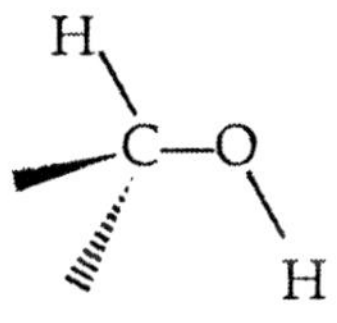

Methanol

Until the late 1930s, methanol was the most widely used antifreeze. While effective in preventing the coolant from freezing, its low boiling point and low specific heat capacity led to considerably less cooling than water alone. Also, the concentration of methanol would tend to be reduced over time due to its greater tendency to evaporate than the water with which it was mixed.

Ethylene Glycol

Ethylene glycol solutions became available in 1937 and were marketed as "permanent antifreeze", since the higher boiling points provided advantages for summertime use as well as during cold weather. They are still used today. Ethylene glycol antifreezes are poisonous and should be kept away from any person or animal (children and especially dogs) that might be tempted by its sweet taste. They form calcium oxalate crystals in the kidneys and can cause acute renal failure and death. All spills should be cleaned, or else an area in which it may be present should be kept inaccessible to those who might ingest it.

Should ingestion of antifreeze occur, ethanol (alcoholic beverages) can be administered until proper treatment can be started in order to slow the conversion of methanol to formaldehyde and formic acid which are the substances responsible for methanol's toxicity. In practice, ethanol can be administered intravenously by doctors to counter ethylene glycol and methanol poisoning, but now that another antidote is available (fomepizole), it is used less and less for the purpose.

In order to prevent ingestion, bittering agent (denatonium benzoate) is usually added to engine coolant to make it taste unpleasant. In the United States, there is legislation before Congress (H.R.2567/S.1110) that would make the use of a bittering agent mandatory.

Propylene Glycol

Propylene glycol, on the other hand, is considerably less toxic and may be labelled as "non-toxic antifreeze". It is used as antifreeze where ethylene glycol would be inappropriate, such as in food-processing systems or in pipes in homes, as well as numerous other settings. It is also used in food, medicines, and cosmetics, often as a binding agent. Propylene glycol is "generally recognised as safe" by the Food and Drug

Administration (FDA) for use in food. However, propylene glycol-based antifreeze should not be considered safe for consumption. In the event of accidental ingestion, emergency medical services should be contacted.

Other Developments

In the 1980s inventor Jack Evans discovered the advantages of using a waterless coolant. His final formulation is a mixture of ethylene glycol and propylene glycol. This coolant has a high boiling point of 370°F and is not corrosive, solving many of water's problems including freezing.

Most commercial antifreeze formulations include corrosion inhibiting compounds, and a coloured dye (commonly a green, red or blue fluorescent) to aid in identification. A 1:1 dilution with water is usually used, resulting in a freezing point of approximately 40°F. In warmer areas weaker dilutions are used.

Glycol antifreeze solutions should generally be replaced with fresh mixture every two years. Many modern cars now come filled with organic acid technology (OAT) antifreeze (such as Dex-Cool, which has an extended service life of five years. Although these still contain glycol, OAT solutions may not be compatible with conventional inorganic-based coolants containing glycol (e.g., with silicates, borates, phosphates) and, if changing from one type to the other, the cooling system must be thoroughly flushed with clean water.

Typically OAT antifreeze contains a red or pink dye to differentiate it from the conventional inorganic coolants (blue or green). Some of the newer technology OAT coolants claim to be compatible with all types of OAT and inorganic-based coolants; these are typically green or yellow in colour.

VAM (Vinyl Acetate)

Vinyl acetate, also known as VAM for vinyl acetate monomer, has the chemical formula $CH_3COOCH{=}CH_2$ and is a colourless liquid with a sweet flavour. Systematic names include 1-acetoxyethylene and acetic acid ethenyl ester.

The usual method of preparation is by reaction of ethylene and acetic acid with oxygen and a palladium catalyst.

It can be polymerised, either by itself to make polyvinyl acetate or with other monomers to prepare copolymers such as ethylene-vinyl acetate.

A relatively reactive monomer due to the instability of the radical derived from VA, attempts to control the polymerisation via most 'living/controlled' radical processes have proved problematic. However, RAFT (or more specifically MADIX) polymerisation offers a convenient method of controlling the synthesis of PVA by the addition of a xanthate chain transfer agent.

8

Photochemical Effects

An example of carbonyl compound which illustrate Photodecarbonylation and example of Abstraction of hydrogen atom to form alcohol.

Cis-trans isomerisation due to absorption of light, illustrated:

Photodecarbonylation

$$(C_6H_5)_2C{=}O + C_6H_5CHOHC_6H_5 \xrightarrow{h\nu} (C_6H_5)_2\underset{|}{\overset{OH}{C}}-\underset{|}{\overset{OH}{C}}(C_6H_5)_2$$

BenzophenoneBenzhydrolBenzpinacol

The photochemical cis trans isomerisation is effected by direct irradiation or in the presence of a sensitiser or a catalyst.

e.g.,

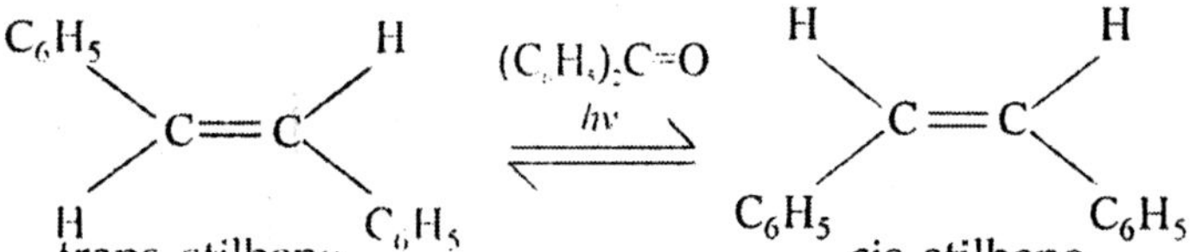

trans stilbene cis stilbene

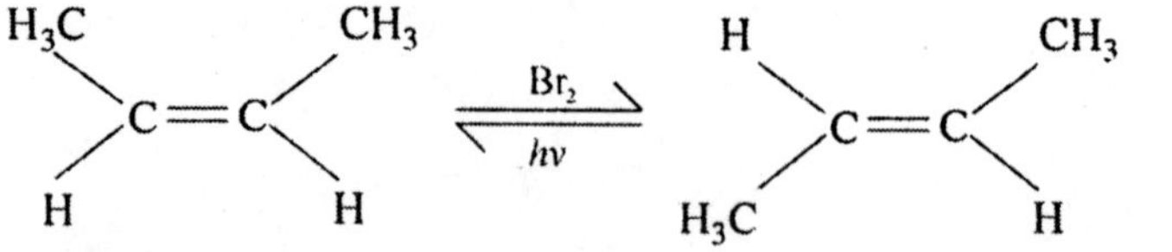

cis-2-butene trans-2-butene

Process of Sesquiterpenoids Caryophyllene

The sesquiterpenoids caryophyllene and its geometrical isomer occur naturally in the oil of cloves. Plan a synthesis of a suitable starting material containing a four membered ring:

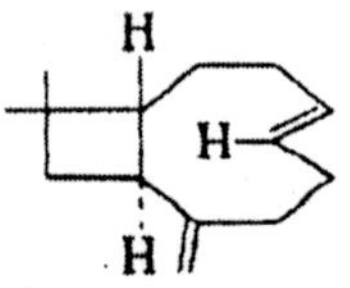

Caryophyllene

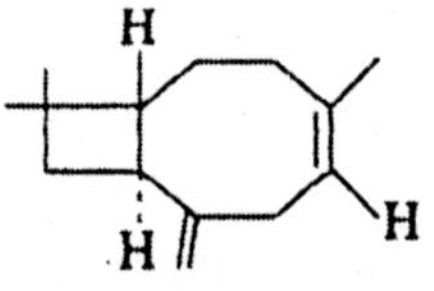

Isocaryophyllene

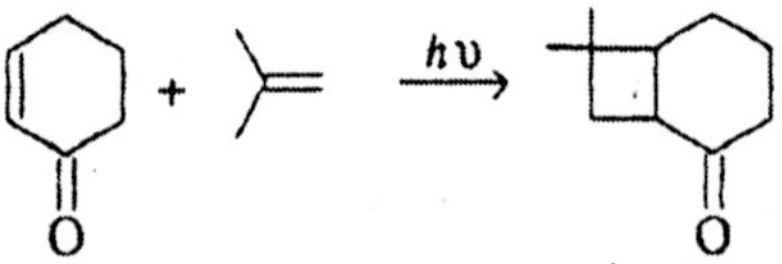

cis + trans

Compounds A and B undergo photodecarboxylation resulting in the formation of I and II (from A) and III and IV (from B). Give their structures and the mechanism of their formations:

(1) 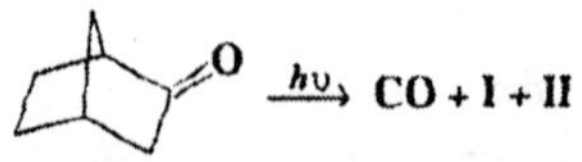

(A)

(2) 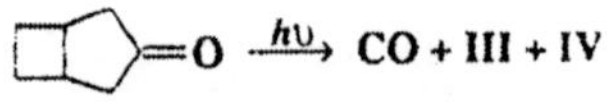

(B)

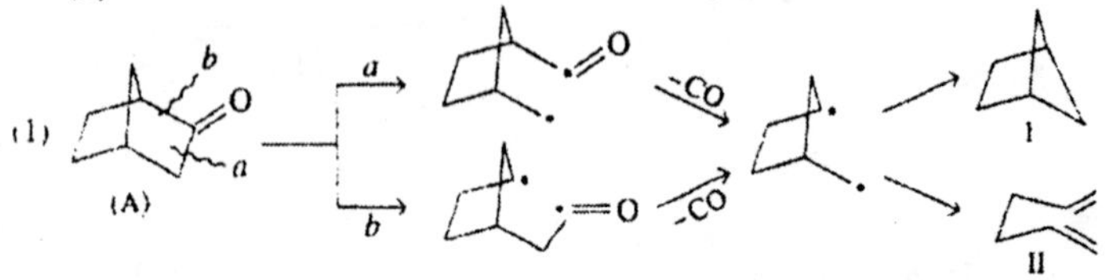

(2) 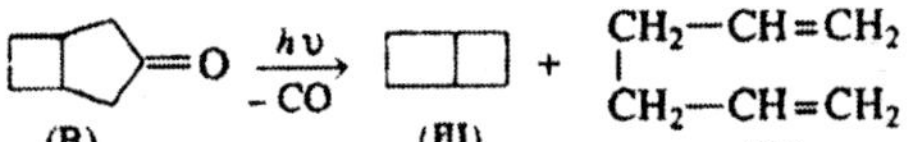

(B) (III) (IV)

The likely products from the vapour phase irradiation of trans 2,6-dimethyl cyclohexanone are :

A mixture of cis and trans-1,2-dimethyl cyclopentanes will be formed as the stereochemical integrity of the molecule would be destroyed on opening of the ring.

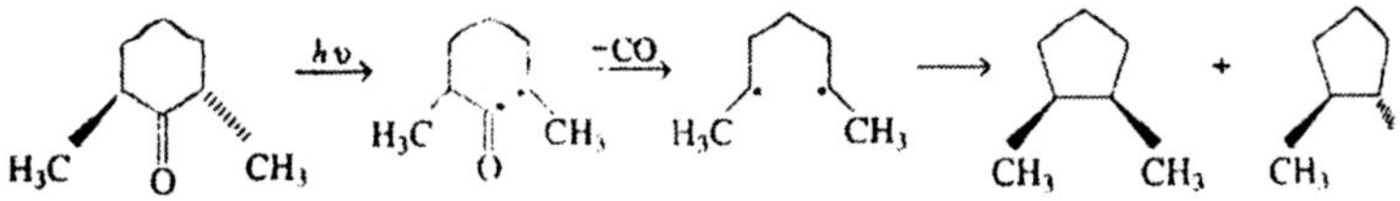

The products you expect to be formed in the irradiation of:

(I) 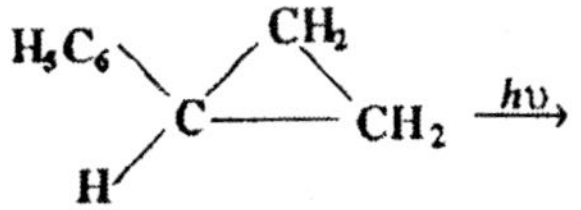

(ii) 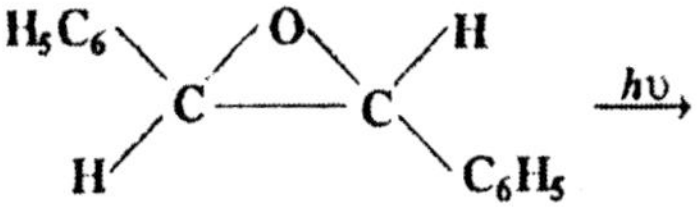

(III) $CH_2=\overset{+}{N}=\overset{-}{N} \xrightarrow{h\nu}$

(*i*) $H_2C: + C_6H_5CH=CH_2$

(*ii*) 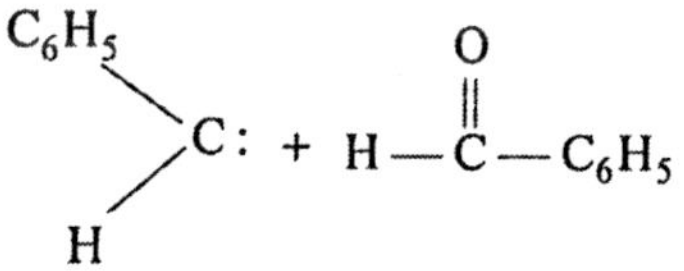

(*iii*) $CH_2 \uparrow\downarrow \longrightarrow CH_2 \uparrow\uparrow$

singlet → triplet

singlet —alkene→ Cyclopropane

triplet —alkene→ Cyclopropane

(Singlet state either decays to triplet and then react with alkene to form cyclopropanes or directly react with alkene to give cyclopropanes)

Suggest mechanism for the following reactions—

(*i*)

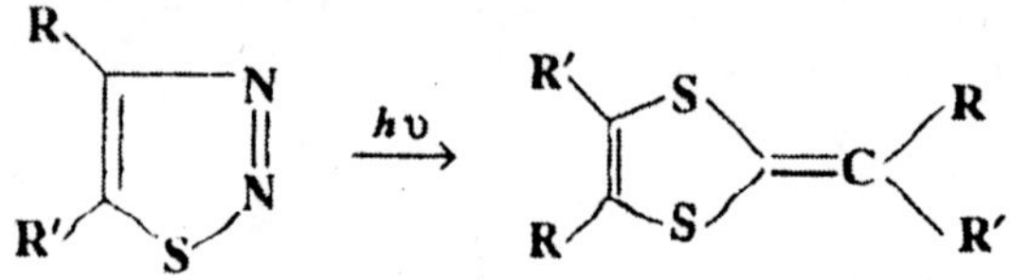

(*ii*)

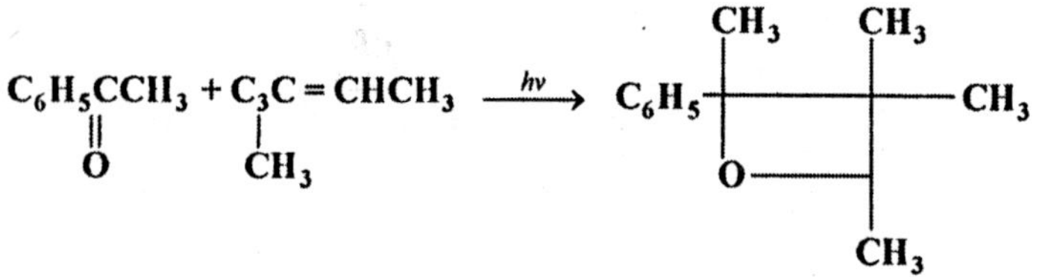

(*i*)

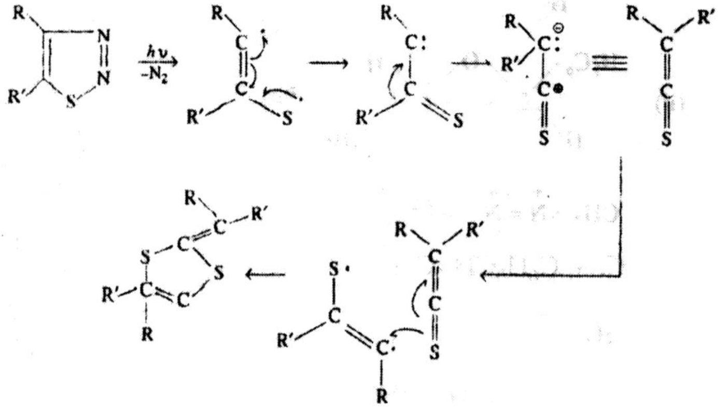

The starting 1,2,3-thiodiazole is formally equivalent to a diazoketone and the first part of the reaction is the same as the photocatalysed Wolf rearrangement.

$$R-\overset{O}{\overset{\|}{C}}\overset{\ominus}{C}H\overset{\oplus}{N}\equiv N \xrightarrow[-N_2]{h\nu} R\underset{O}{\underset{|}{C}}\overset{O}{\overset{\|}{C}}: \rightarrow R\underset{H}{\underset{|}{C}}=C=O$$

(*ii*) 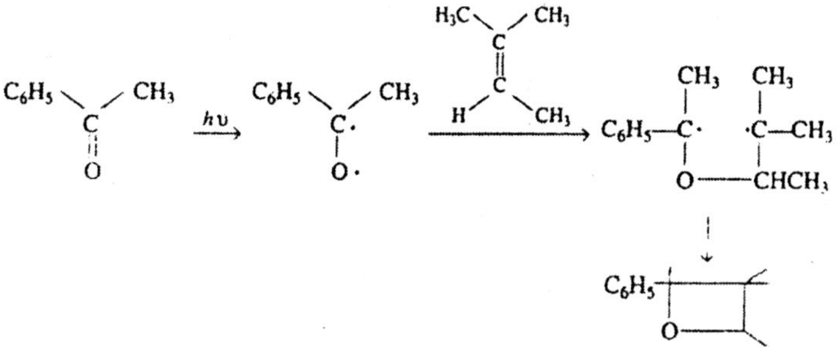

The observed products correspond to the formation of the most stable intermediate by the addition of the photoexcited triplet (diradical) state of acetophenone to the alkene.

The Products Predicted:

(i) 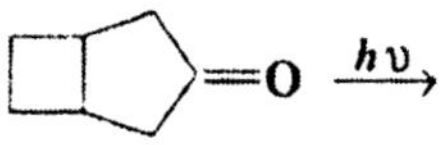

(ii) 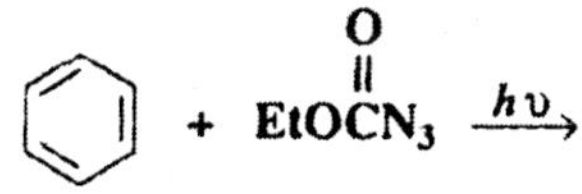

(iii) 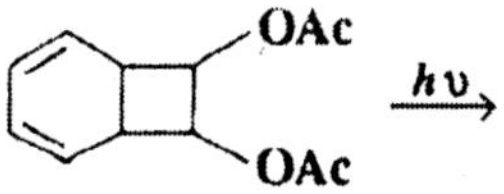

(iv) $Ph_2C = O + (CH_3)_2CHOH \xrightarrow{h\upsilon}$

(i) 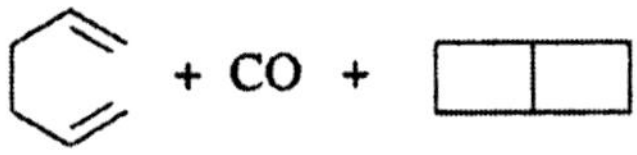

1, 5 hexadiene — Major Product; Bicyclo [2, 2, 0]– hexane — Minor Product

(ii) 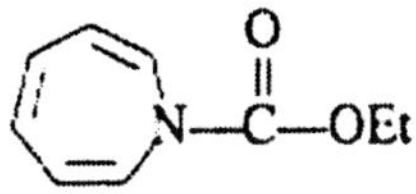

(iii) 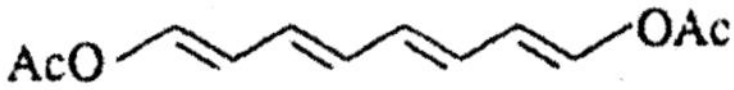

(iv) $Ph_2C(OH)—C(OH)—Ph_2 + CH_3COCH_3$

Benzpinacol

(a) Mentioned two reasons as to why benzophenone is a good triplet sensitiser.

(b) The following compounds will quench benzophenone triplets at a diffusion controlled rate:

(*a*) Sensitisers should be chosen to produce high yields of triplets for maximum efficiency, benzophenone is a good example of this category because—

(*i*) benzophenone forms triplets when irradiated in fluid solution at room temperature;

(*ii*) benzophenone can efficiently transfer its triplet energy to the triplet state of other molecules. Its inter system crossing efficiency is almost equal to 1.

(*b*) Benzophenone has E_T = 69. Thus, butadiene (E_T = 60), anthracene (E_T = 42) and 2-acetonaphthone (E_T = 59) will quench benzophenone triplets at the diffusion control rat. Whereas benzene (E_T = 85) and acetophenone (E_T = 75) will not.

The following compound is expected or not to undergo efficient photoreduction:

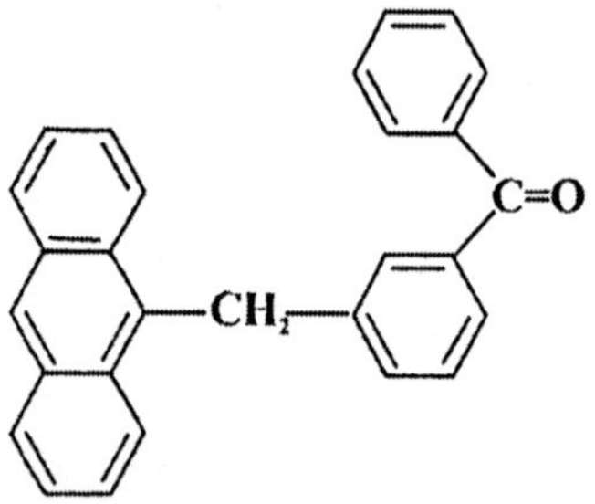

No, because the molecule carries a built in quencher around with it. Each benzophenone triplet formed will be quenched at a rate much faster than hydrogen abstraction from solvent.

The major products from the photosensitised dimerisation of isoprene predicted. The ratio of diene dimers vary as the sensitiser is varied or not:

(1) 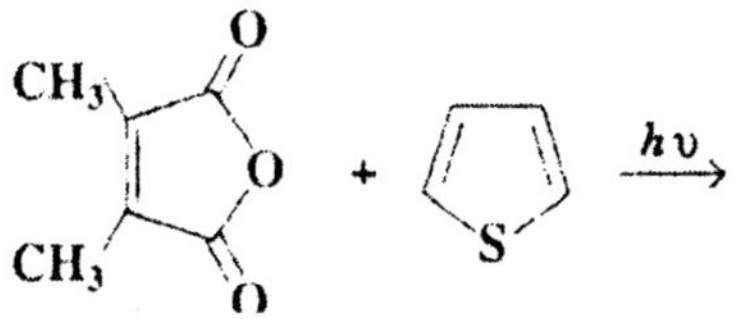

(2) 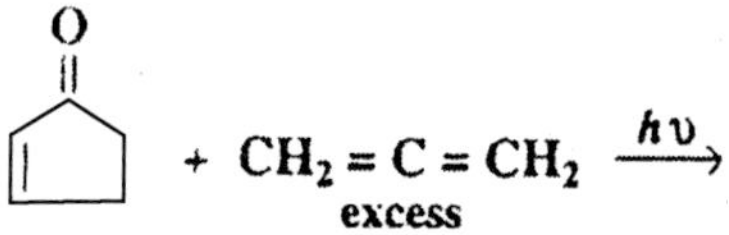

(*a*) 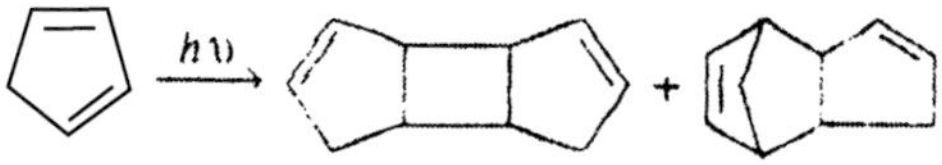

The ratio of dimers will not vary because stereoisomeric triplets are not possible.

(1) 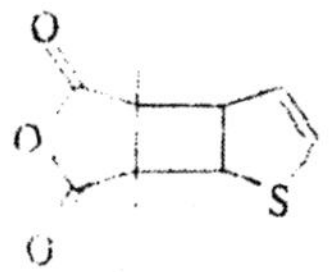

(2) 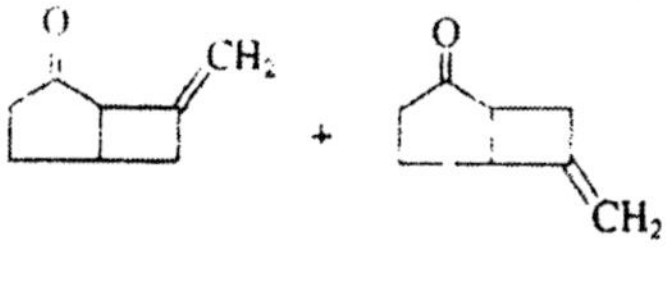

Process of Intramolecular Hydrogen Abstraction

Intramolecular hydrogen abstraction from a γ carbon occurs when certain ketones are irradiated. For, *e.g.*, vapours of Hexan-2-one on photolysis gives an alkene and a ketone (via the enol form). It is referred to as Norrish type II cleavage.

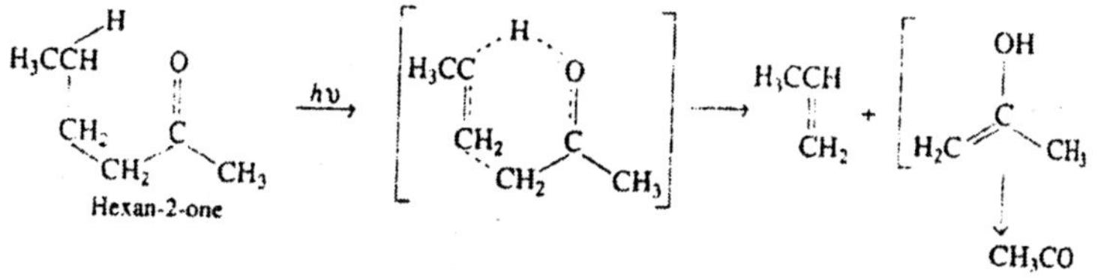

Matched:

Reaction	***Type of Photoisomerisation reaction***
(*a*) *cis* $ML_nX_m \xrightarrow{h\nu}$ *trans* ML_nX_m $n = 2$ or 4; $m = 2$ $n = 3$; $m = 3$	*Optical*
(*b*) $(d)M(L—L)_3 \xrightarrow{h\nu} (l)M(L—L)_3$	*Linkage*
(*c*) $L_nM—XY \xrightarrow{h\nu} L_nM—Y—X$	*Geometrical*

(*a*) Geometrical,

(*b*) Optical,

(*c*) Linkage.

The reactions completed:

(a) 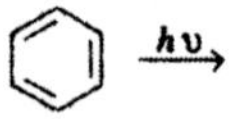

(b) 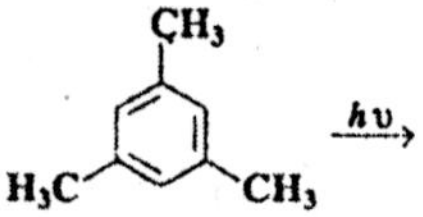

(c) 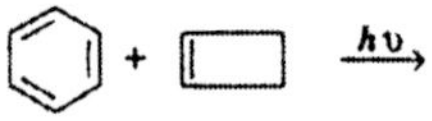

(d) 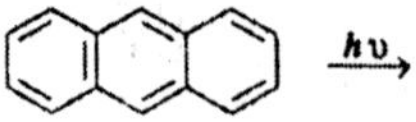

(e) 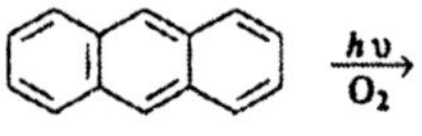

(f) 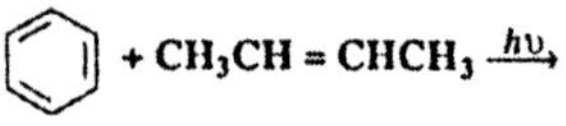

(a) 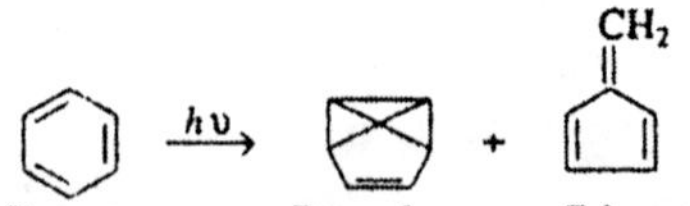

Benzene Benzyalene Fulvene

(b) 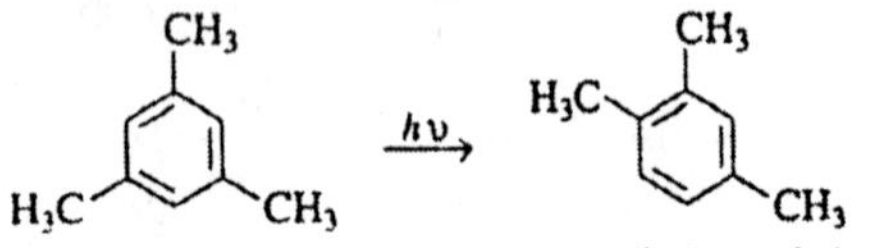

1, 3, 5 trimethyl benzene 1, 2, 4 trimethyl benzene

(Isomerisation)

(c) 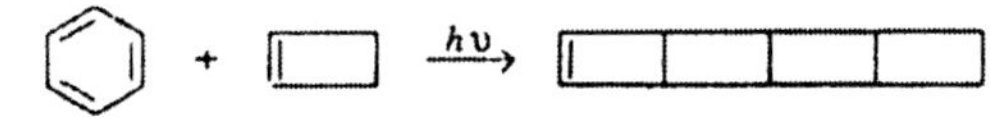

Benzene Cyclobutene Tetracyclic ene

(*d*) 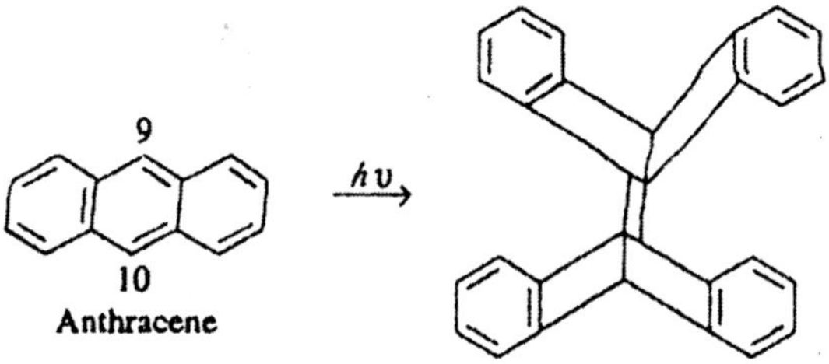

(Dimerisation)
Join at 9, 10 position

(*e*) 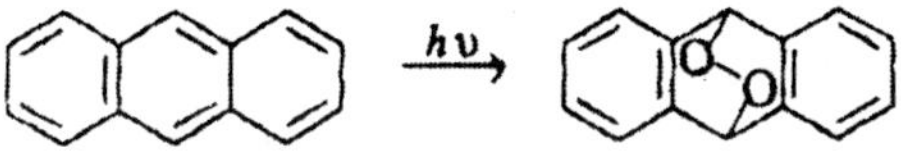

Anthracene Transanular (9, 10) peroxide of Anthracene

(*f*) 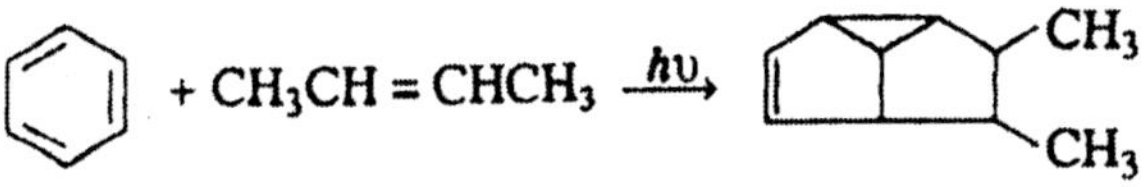

(a) Drawn and labelled Jablonski diagram:

(b) The product of the following reaction:

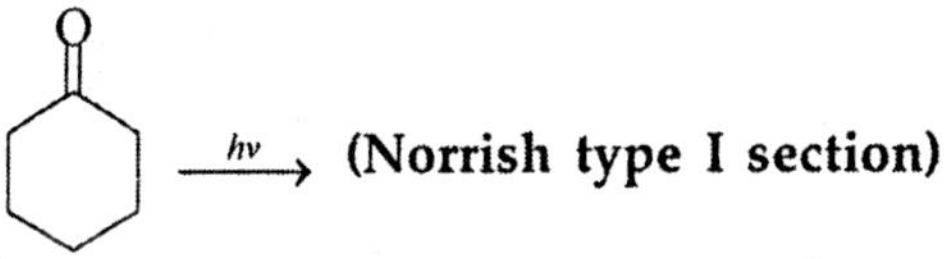

(*a*) 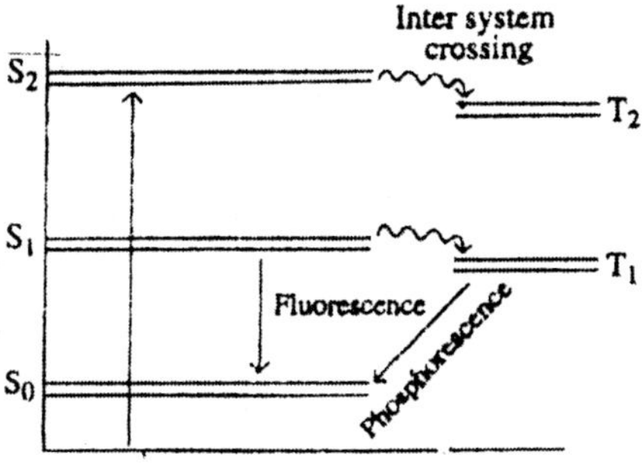

(Jablonski Diagram)

(*b*) 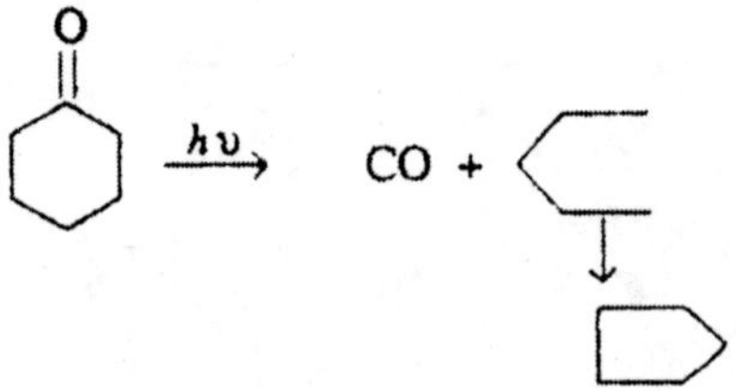

Drawn diagram for the relative energies of the various types of π orbitals of butadiene (which has two bonding and two antibonding π orbitals) to illustrate the ground state and the S_1, T_1, S_2 and T_2 excited states.

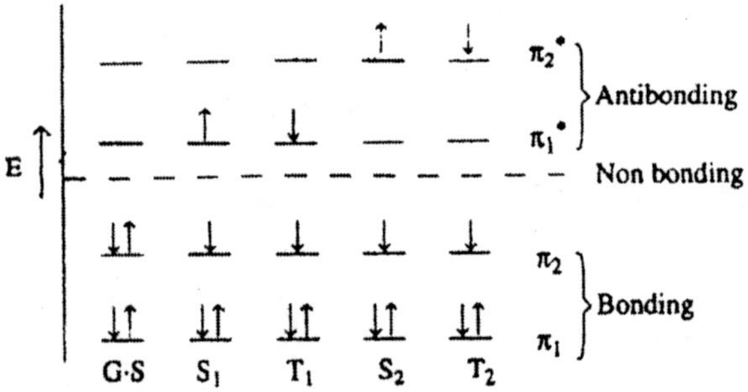

The following equations completed:

(a) 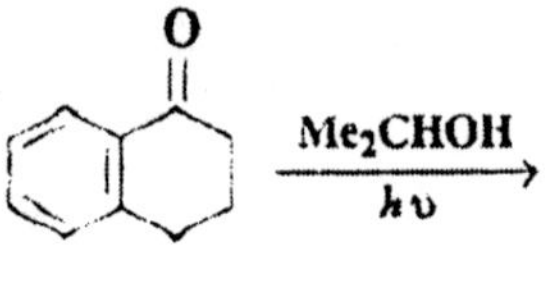

(b) $Me_2CHONO \xrightarrow{h\upsilon}$

(c) 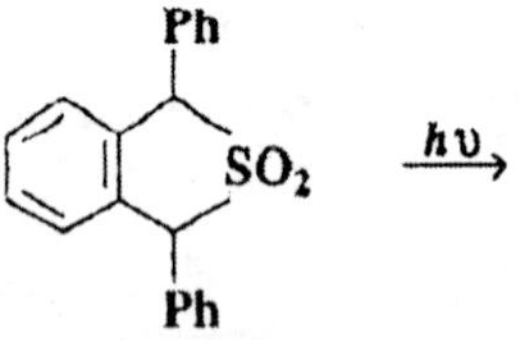

(d) 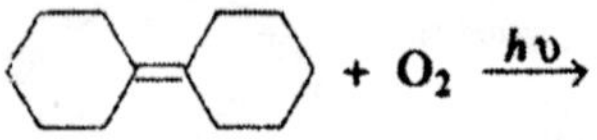

(e) $MeCH_2CH_2COMe \xrightarrow{h\nu}$ (a cyclobutane is m:jor product)

(f) [cyclodecanone] $\xrightarrow{h\nu}$ **(a decalin)**

(g) [2-azidobiphenyl, N_3] $\xrightarrow{h\nu}$ **(a heterocycle)**

(h) $CH_3CH_2CH_2CH_2N_3 \xrightarrow{h\nu}$ **(a cyclic compound)**

(a) [1-tetralone] + $Me_2CHOH \xrightarrow{h\nu}$ [1-hydroxytetralin radical, OH] + $M_2\dot{C}—OH$

2 [OH radical] $\longrightarrow$ [dimer, HO ... OH]

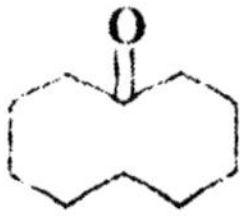

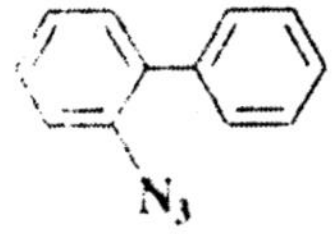

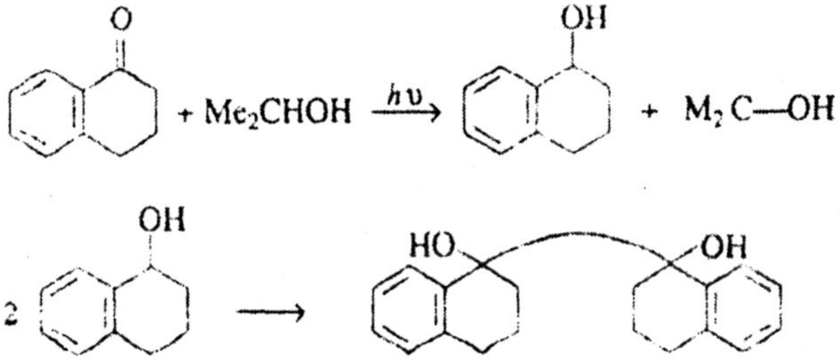

(*b*) $Me_2CHONO \xrightarrow{h\nu} Me_2CHO + NO$

$Me_2C—O\bullet$ / H $\bullet OCHMe_2 \longrightarrow Me_2C{=}O + HOCHMe_2$

(*c*) [Ph, SO_2, Ph] $\xrightarrow{h\nu}$ SO_2 + [Ph, Ph diradical] $\longrightarrow$ [Ph, Ph] trans

(d) [H] + O_2 $\xrightarrow[\text{Sensitized}]{h\nu}$ [H, O, O] $\longrightarrow$ [O, H—O]

(e) [H, O, CH_2 CMe, CH_2—CH_2] $\xrightarrow{h\nu}$ [HO Me] $\longrightarrow$ [OH, Me]

(f) [O] $\xrightarrow{h\nu}$ [$\dot{O}$, H, H] $\longrightarrow$ [OH, H] $\longrightarrow$ [OH, H] cis

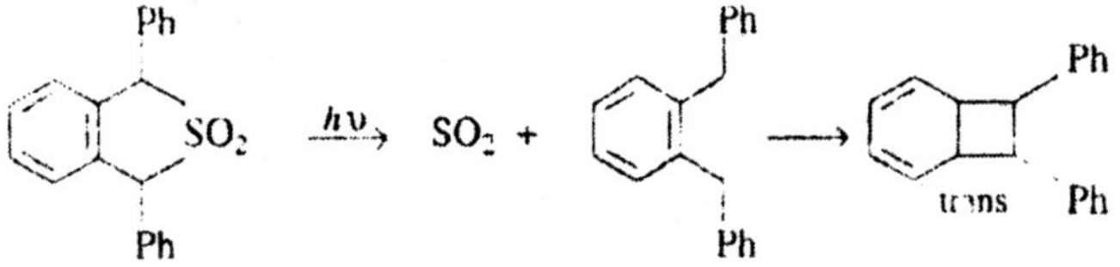

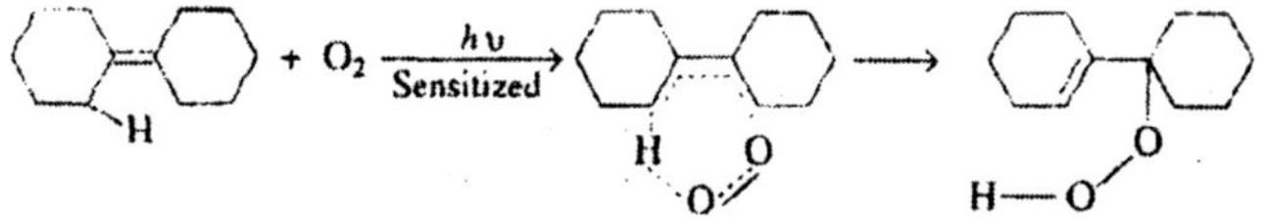

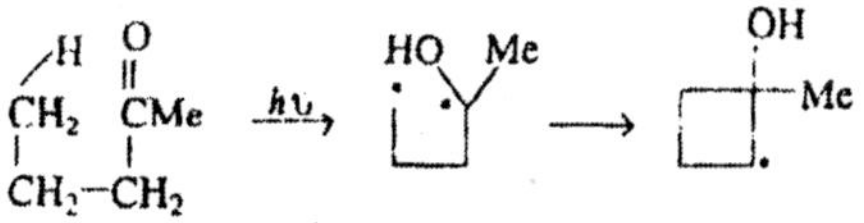

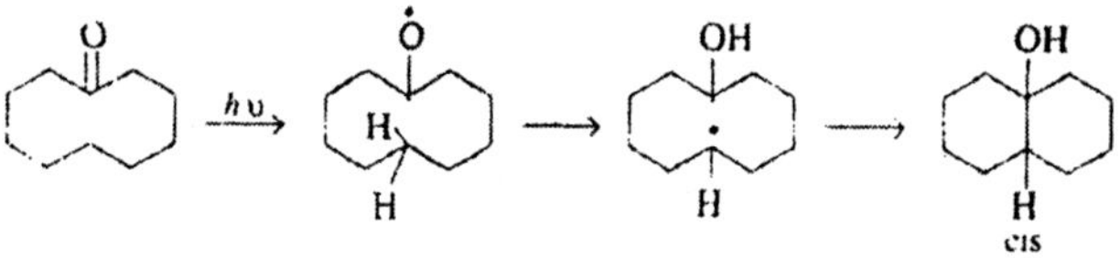

(g)

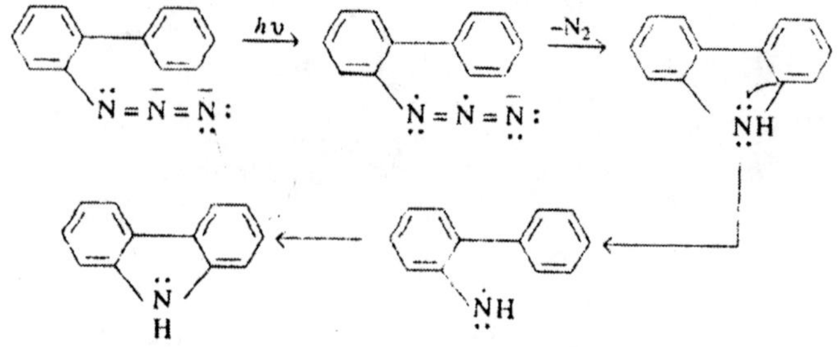

(h)

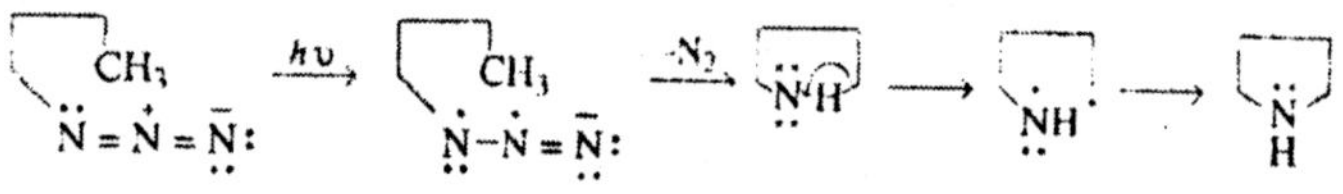

A mechanism for the reaction suggested:

$$\mathbf{2PhCH_2CH_2ONO \xrightarrow{h\nu} (PHCH_2NO)_2 + 2CH_2O}$$

$$PhCH_2—CH_2O—NO \xrightarrow{h\nu} NO + PhCH_2—CH_2O. \longrightarrow CH_2O + NO + PhCH_2.$$

$$\text{Dimer} \longleftarrow PhCH_2NO \longleftarrow$$

The Norrish type I process is not important for the photolysis of diaryl ketones. A reason suggested:

Norrish type I process is—

$$R^1CO—R^2 \xrightarrow{h\nu} R^{\cdot} + R^2—\dot{C}O \longrightarrow R^{2\bullet} + CO$$

If diaryl ketones followed this route we would have

$$Ar^1—CO—Ar^2 \xrightarrow{h\nu} Ar^{1\bullet} + Ar^2—\dot{C}O \longrightarrow Ar^{2\bullet} + CO$$

Now aryl free radicals are extremely unstable, they are not stabilised by resonance. Since the stability of a product can act as a driving force for the reaction to proceed along that path then we can say that the more unstable the product is, the less likely that path will be followed.

Hence Norrish type I process is energetically unfavourable for diaryl ketones.

(a) A synthesis of the following compound, planned:

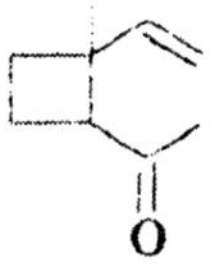

(b) The structures of the photosensitised irradiation of the following compounds:

(I)

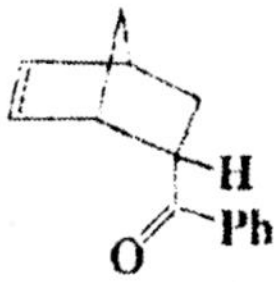

(II)

(c) Rationalise the following reaction:

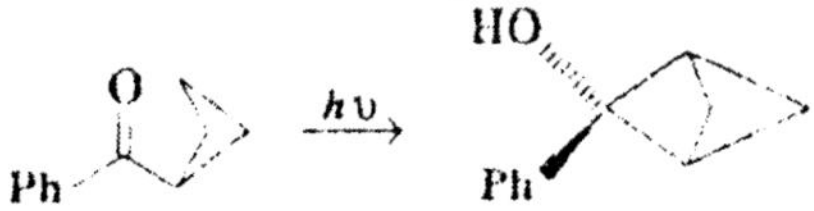

(I) (II)

(*a*) The cyclobutane ring can be constructed via photochemical cycloaddition of the alkene using acetophenone sensitiser.

(*b*) (I)  (II) 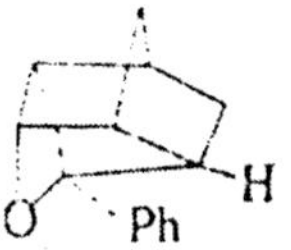

(*c*) The photochemical conversion of I to II involves the coupling rather than the fragmentation of the Norrish type II biradical to the cyclobutanol.

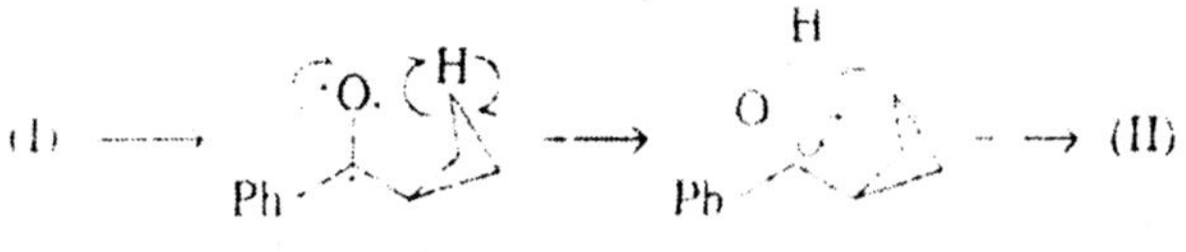

(n, π^*)

1,4-cineole and ascaridole are the naturally occurring oxide and peroxide respectively. There are three isomeric terpinenes.

The following out of these is a suitable isomer for photochemical conversion into either 1,4-cineole or ascaridole:

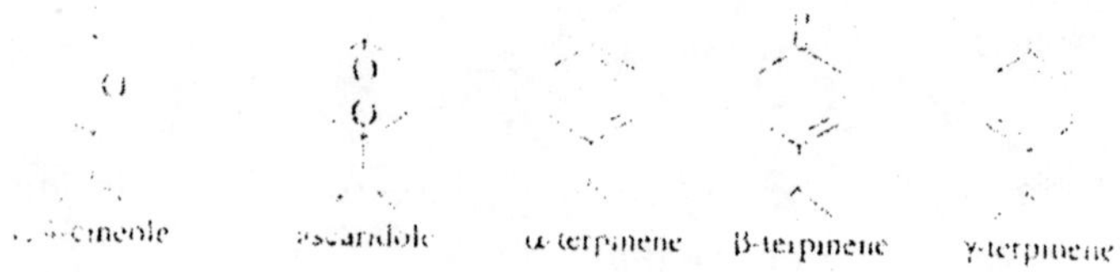

Formation of cyclic peroxides by conjugated dienes is a general reaction and although ultraviolet light often initiates the reaction, better results are achieved by carrying out the irradiation of α-terpinene in the presence of sensitisers, *e.g.*, chlorophyll, dyes etc.

Bibliography

Adams, H.R.: *Veterinary Pharmacology and Therapeutics*, Iowa State Univ. Press, Ames, 1995.

Adams, R.D.: *Principles of Neurology*, McGraw Hill, New York, 1997.

Arnold, H.L.: *Andrews' Diseases of the Skin*, Saunders, Philadelphia, 1990.

Barclay, A.N.: *The Leucocyte Antigen Factsbook*, Academic Press, San Diego, 1997.

Behrman, R.E.: *Nelson Textbook of Pediatrics*, Saunders, Philadelphia, 2004.

Belshe, R.B.: *Textbook of Human Virology*, Mosby, St. Louis, 1991.

Berman, H.J.: *The Financial Management of Hospitals*, Health Administration Press, Ann Arbor, 1994.

Blauvelt, C.T. and Nelson F.R.T.: *A Manual of Orthopaedic Terminology*, Mosby, St. Louis, 1994.

————: *Current Procedural Terminology: CPT 2006*, AMA Press, Chicago, 2005.

Bradley, W.G.: *Neurology in Clinical Practice*, Butterworth Heinemann, Boston, 2000.

Brenner, D.J. Krieg N.R. and Staley J.T.: *Bergey's Manual of Systematic Bacteriology*, Springer, New York, 2005.

Brunton, L.L.: *Goodman and Gilman's the Pharmacological Basis of Therapeutics*, McGraw-Hill, New York, 2006.

Brusca, R. C. and Brusca G.J.: *Invertebrates*, Sinauer Associates, Sunderland, 2002.

Campbell, R.J.: *Campbell's Psychiatric Dictionary*, Oxford Univ. Press, New York, 2004.

Cassileth, B. R.: *The Alternative Medicine Handbook*, Norton Co., New York, 1998.

Clemente, C.D.: *Federative Committee on Anatomical Terminology*, Terminologica Anatomica, Stuttgart, 1998.

————————: *Gray's Anatomy*, Lea & Febiger, Philadelphia, 1985.

Cruse, J. M. and Lewis R.E.: *Illustrated Dictionary of Immunology*, Taylor and Francis, Boca Raton, 2003.

Cunningham, F.G.: *Williams Obstetrics*, Appleton & Lange, Stamford, 1997.

Dale, D.C.: *Dorland's Illustrated Medical Dictionary*, Saunders, Philadelphia, 2003.

————————: *Scientific American Medicine*, Scientific American, New York, 2004.

DeVita, V.T.: *Cancer: Principles and Practice of Oncology*, Lippincott, Philadelphia, 2005.

Donaghy, M.: *Brain's Diseases of the Nervous System*, Oxford, New York, 2001.

Ekholm, P.: *Effects of Selenium Supplemented Commercial Fertilizers on Food Selenium Content and Selenium Intake in Finland*, University of Helsinki, 1997.

Fishman, A.P.: *Fishman's Pulmonary Diseases and Disorders*, McGraw Hill, New York, 1998.

Fitzgerald, K.A.: *The Cytokine Factsbook*, Academic Press, San Diego, 2001.

Fritz, A: *International Classification of Diseases for Oncology (ICD O)*, World Health Organization, Geneva, 2000.

Gallagher, L.A.: *Thesaurus of Psychological Index Terms*, American Psychological Association, Washington, 2005.

Gennaro, A. R.: *Hawley's Condensed Chemical Dictionary*, John Wiley, New York, 2001.

Guyton, A.C. and Hall J.E.: *Textbook of Medical Physiology*, Saunders, Philadelphia, 2006.

Haubrich, W.S.: *Bockus Gastroenterology*, Saunders, Philadelphia, 1995.

Herbert, W.J.: *Dictionary of Immunology*, Academic Press, London, 1995.

Hesketh, R.: *The Oncogene and Tumour Suppressor Core Factsbook*, Academic Press, San Diego, 1997.

Jablonski, S.: *Jablonski's Dictionary of Syndromes and Eponymic Diseases*, Krieger, Malabar, 1991.

——————: *McGraw Hill Dictionary of Scientific and Technical Terms*, McGraw Hill, New York, 2003.

Jameson, J. L.: *Principles of Molecular Medicine*, Humana Press, Totowa, 1998.

Kasper, D.L.: *Harrison's Principles of Internal Medicine*, McGraw Hill, New York, 2005.

Kendrew, J.C. and Lawrence E.: *The Encyclopedia of Molecular Biology*, Blackwell Science, Cambridge, 1994.

King, R.C. and Stansfield W.D.: *A Dictionary of Genetics*, Oxford Univ. Press, New York, 2002.

Kirk, P. M.: *Ainsworth and Bisby's Dictionary of the Fungi*, CAB International, England, 2001.

Lackie J.M and Dow J.A.T.: *Dictionary of Cell and Molecular Biology*, Academic Press, San Diego, 1999.

Larsen, P.R.: *Williams Textbook of Endocrinology*, Saunders, Philadelphia, 2003.

Leigh, G. J.: *Principles of Chemical Nomenclature*, Blackwell Science, Cambridge, 1998.

Lide, D.R.: *CRC Handbook of Chemistry and Physics*, CRC Press, Boca Raton, 1991.

Markell, E.K.: *Medical Parasitology*, Saunders, Philadelphia, 1992.

Menkes, J.H.: *Textbook of Child Neurology*, Williams & Wilkins, Baltimore, 1995.

Metzler, D. E.: *Biochemistry: The Chemical Reactions of Living Cells*, Harcourt / Academic Press, San Diego, 2001.

Migloni, G.S.: *Dictionary of Plant Venetics and Molecular Biology*, Hawthorne Press, New York, 1998.

Milne, G.W.: *Drugs: Synonyms and Properties*, Ashgate, Hampshire, 2000.

Moore, K.L.: *Clinically Oriented Anatomy*, Williams & Wilkins, Baltimore, 1992.

Moschella, S.L. and Hurley H.J.: *Dermatology*, Saunders, Philadelphia, 1992.

Negwer, M.: *Organic Chemical Drugs and Their Synonyms*, Akademie Verlag, Berlin, 1994.

Nill, K.: *Glossary of Biotechnology and Nanobiotechnology Terms*, CRC Press, 2006.

O'Neil, M.: *The Merck Index: An Encyclopedia of Chemicals, Drugs and Biologicals*, Merck, New Jersey, 2001.

Parfitt, K.: *Martindale: The Complete Drug Reference*, Pharmaceutical Press, London, 1999.

Pigott, R. and Power C: *The Adhesion Molecule Factsbook*, Academic Press, San Diego, 1993.

Scott, T.A.: *Concise Encyclopedia of Biology*, W. De Gruyter, Berlin, 1996.

Slee, V.N. and Slee, D.A.: *Health Care Reform Terms*, Tringa Press, St. Paul, 1993.

——————: *Health Care Terms*, Tringa Press, St. Paul, 1996.

Sloane, S.B.: *Medical Abbreviations and Eponyms*, Saunders, Philadelphia, 1997.

Stetzenbach, L. D and Yates M. V.: *Dictionary of Environmental Microbiology*, Academic Press, San Diego, 2003.

Trissel, L. A.: *Handbook on Injectable Drugs*, American Society of Health-System Pharmacists, Bethesda, 2005.

Williams, S.T.: *Bergey's Manual of Systematic Bacteriology*, Williams & Wilkins, Baltimore, 1989.

Yanoff, M. and Duker J. S.: *Ophthalmology*, Mosby, St. Louis, 2004.

Zipes, O.P.: *Braunwald's Heart Disease*, Elsevier-Saunders, Philadelphia, 2005.

Index

□□□